Marine Protected Areas

TOOLS FOR SUSTAINING
OCEAN ECOSYSTEMS

Committee on the Evaluation, Design, and Monitoring of Marine Reserves and
Protected Areas in the United States
Ocean Studies Board
Commission on Geosciences, Environment, and Resources
National Research Council

NATIONAL ACADEMY PRESS
Washington, D.C.

NATIONAL ACADEMY PRESS • 2101 Constitution Ave., N.W. • Washington, DC 20418

NOTICE: The project that is the subject of this report was approved by the Governing Board of the National Research Council, whose members are drawn from the councils of the National Academy of Sciences, the National Academy of Engineering, and the Institute of Medicine. The members of the committee responsible for the report were chosen for their special competencies and with regard for appropriate balance.

This report and the committee were supported by grants from the National Oceanic and Atmospheric Administration, the National Park Service, and the U.S. Fish and Wildlife Service. The views expressed herein are those of the authors and do not necessarily reflect the views of the sponsors.

Library of Congress Cataloging-in-Publication Data

Marine protected areas : tools for sustaining ocean ecosystems /
Committee on the Evaluation, Design, and Monitoring of Marine Reserves
and Protected Areas in the United States Ocean Studies Board
Commission on Geosciences, Environment, and Resources National Research
Council.
    p. cm.
Includes bibliographical references (p.   ).
 ISBN 0-309-07286-7 (hard)
 1. Marine parks and reserves. I. National Research Council (U.S.).
Committee on the Evaluation, Design, and Monitoring Marine Reserves and
Protected Areas in the United States. II. Title.
 QH91.75.A1 M28 2001
 333.78'4—dc21
                                     2001000995

Marine Protected Areas: Tools for Sustaining Ocean Ecosystems is available from the National Academy Press, 2101 Constitution Avenue, N.W., Box 285, Washington, DC 20055; (800) 624-6242 or (202) 334-3313 (in the Washington Metropolitan area); Internet: http://www.nap.edu

THE NATIONAL ACADEMIES

National Academy of Sciences
National Academy of Engineering
Institute of Medicine
National Research Council

The **National Academy of Sciences** is a private, nonprofit, self-perpetuating society of distinguished scholars engaged in scientific and engineering research, dedicated to the furtherance of science and technology and to their use for the general welfare. Upon the authority of the charter granted to it by the Congress in 1863, the Academy has a mandate that requires it to advise the federal government on scientific and technical matters. Dr. Bruce M. Alberts is president of the National Academy of Sciences.

The **National Academy of Engineering** was established in 1964, under the charter of the National Academy of Sciences, as a parallel organization of outstanding engineers. It is autonomous in its administration and in the selection of its members, sharing with the National Academy of Sciences the responsibility for advising the federal government. The National Academy of Engineering also sponsors engineering programs aimed at meeting national needs, encourages education and research, and recognizes the superior achievements of engineers. Dr. William A. Wulf is president of the National Academy of Engineering.

The **Institute of Medicine** was established in 1970 by the National Academy of Sciences to secure the services of eminent members of appropriate professions in the examination of policy matters pertaining to the health of the public. The Institute acts under the responsibility given to the National Academy of Sciences by its congressional charter to be an adviser to the federal government and, upon its own initiative, to identify issues of medical care, research, and education. Dr. Kenneth I. Shine is president of the Institute of Medicine.

The **National Research Council** was organized by the National Academy of Sciences in 1916 to associate the broad community of science and technology with the Academy's purposes of furthering knowledge and advising the federal government. Functioning in accordance with general policies determined by the Academy, the Council has become the principal operating agency of both the National Academy of Sciences and the National Academy of Engineering in providing services to the government, the public, and the scientific and engineering communities. The Council is administered jointly by both Academies and the Institute of Medicine. Dr. Bruce M. Alberts and Dr. William A. Wulf are chairman and vice chairman, respectively, of the National Research Council.

v

 *Foreword*

The Ocean Studies Board (OSB) is pleased to present this report, *Marine Protected Areas: Tools for Sustaining Ocean Ecosystems*. It represents the culmination of a two-year, in-depth examination of this controversial approach to marine resource management that required analysis of issues in both marine ecology and fisheries science.

For many years the OSB has been interested in topics concerning marine ecology and the preservation of marine biodiversity. Notable reports in this area include *Priorities for Coastal Ecosystem Science* (1994), *Understanding Marine Biodiversity* (1995), and *From Monsoons to Microbes: Understanding the Ocean's Role in Human Health* (1999). At the same time, the board has concerned itself with the sound, science-based management of marine fisheries, as exemplified by studies such as *Improving Fish Stock Assessments* (1998), *Sharing the Fish: Toward a National Policy on Individual Fishing Quotas* (1999), and *Sustaining Marine Fisheries* (1999). These two interests come together on the issue of marine reserves, which have been proposed as an ecosystem-based approach for conserving living marine resources, both for fisheries management and for preserving marine biodiversity.

It is our hope that this report will serve as a sound basis for future efforts to design and implement marine reserves and protected areas. It provides a summary of what we know, recommendations about how to apply that knowledge, and a description of what we need to know to maximize the effectiveness of this marine management tool.

The board is grateful to the committee members who volunteered enormous amounts of their time to complete this ambitious undertaking.[*]

Kenneth Brink
Chair, Ocean Studies Board

---

[*] To view this report on-line, or to learn more about the OSB's mission and other projects, please visit our Web site at www.national-academies.org/osb.

Preface

The concept of marine reserves has been repeatedly addressed in the past 25 years, but implementation and subsequent evaluation of these protected areas has been relatively infrequent until the past decade. In recent years, there has been strong advocacy for reserves among the conservation community and those concerned about losses of habitat and biodiversity in the sea. At the same time, conventional users of marine resources, especially fishing industries and communities, have asked serious questions about the efficacy of marine reserves as a tool for resource management because of the modest level of experience with their proper design, siting, and evaluation. The Ocean Studies Board appointed a committee with broad disciplinary expertise to objectively investigate the potential use of marine reserves with respect to design, implementation criteria, and probable efficacy in relation to meeting biodiversity, conservation, and fisheries management goals. Issues emphasizing ecology, oceanography, and socioeconomic impacts are prominent in the report, which strives to integrate and synthesize the diverse information on reserves, followed by conclusions and recommendations.

Few would deny that the oceans are stressed by human activities and that new, or additional, management measures are required to ensure that the ocean's living resources and ecosystem services are conserved. The concept of designating specific areas as marine protected areas (MPAs) and reserves proffers another tool with the potential for expanding our ability to manage resources. Increasing designation and implementation of reserves represent a shift in emphasis toward spatially explicit management measures, an emphasis that many believe is needed given the present heavy utilization of ocean resources. The recent presidential executive order (May 2000) directing the Department of Commerce and the

Department of the Interior to develop a plan for MPA networks in U.S. coastal waters is one major step toward wider application of this approach. This report will serve as a comprehensive and critical description and evaluation of MPAs and reserves as a management tool that can help to guide agencies as they move forward in developing plans for a national system of MPAs.

The Committee on the Evaluation, Design, and Monitoring of Marine Reserves and Protected Areas is very grateful to the many individuals who played a significant role in the completion of this study. The committee met five times and would like to extend its gratitude to all of the individuals who appeared before the full committee or otherwise provided background information and discussed pertinent issues (see Appendix D for a complete list of speakers and participants).

This report has been reviewed in draft form by individuals chosen for their diverse perspectives and technical expertise, in accordance with procedures approved by the National Research Council's (NRC's) Report Review Committee. The purpose of this independent review is to provide candid and critical comments that will assist the institution in making the published report as sound as possible and to ensure that the report meets institutional standards for objectivity, evidence, and responsiveness to the study charge. The review comments and draft manuscript remain confidential to protect the integrity of the deliberative process. We wish to thank the following individuals for their review of this report: Tundi Agardy (Conservation International), Ann Bucklin (University of New Hampshire), Larry Crowder (Duke University Marine Laboratory), Christopher D'Elia (State University of New York at Stony Brook), Paul Durrenberger (Pennsylvania State University), Jane Lubchenco (Oregon State University), James MacMahon (Utah State University), Melissa Miller-Henson (California Resources Agency), and Richard Young (commercial fisherman). Although the reviewers listed above provided many constructive comments and suggestions, they were not asked to endorse the conclusions and recommendations nor did they see the final draft of the report before its release. The review of this report was overseen by H. Ronald Pulliam (University of Georgia), appointed by the Commission on Geosciences, Environment, and Resources and Robert Frosch (Harvard University), appointed by the NRC's Report Review Committee, who were responsible for making certain that an independent examination of the report was carried out in accordance with institutional procedures and that all review comments were carefully considered. Responsibility for the final content of this report rests entirely with the authoring committee and the institution.

The committee extends its thanks to the staff of the Ocean Studies Board (OSB) of the National Research Council (NRC), who provided both leadership and logistical support for the study. Study Director Susan Roberts tirelessly contributed her time to all aspects of the study, and her important contributions to the study and report are gratefully acknowledged. Senior Project Assistant Ann Carlisle provided superb logistical support throughout the study and during report preparation. OSB Director Morgan Gopnik and OSB Senior Program Officer, Ed Urban, both provided critical comments and editorial advice during the preparation of the report. Merrie Cartwright and Kate Shafer provided valuable research assistance during their internships at the NRC. Additionally, Associate Director of the Board on Environmental Studies and Toxicology David Policansky, participated in several committee meetings and contributed valuable ideas and expertise.

The committee is also grateful for the assistance provided by the following individuals who provided additional background material, data, publication lists, and figures for consideration and use by the committee: Bill Ballantine (Leigh Marine Laboratory, New Zealand), Jim Bohnsack (National Marine Fisheries Service), Elizabeth Clarke (National Marine Fisheries Service), Jeff Cross (Sandy Hook Laboratories), Larry Crowder (Duke University Marine Laboratory), Michael Murphy (National Marine Fisheries Service), and Mike Pentony (New England Fishery Management Council). We would also like to thank the many institutions and organizations that provided us with related background information, reference materials, and reports.

Edward Houde
Chair

Contents

Executive Summary

Declining yields in many fisheries and the decay of treasured marine habitats such as coral reefs have heightened interest in establishing a comprehensive system of marine protected areas (MPAs) in the United States. **MPAs, areas designated for special protection to enhance the management of marine resources, show promise as components of an ecosystem-based approach for conserving the ocean's living assets. However, MPA proposals often raise significant controversy, especially the provisions for marine reserves—zones within an MPA where removal or disturbance of resources is prohibited, sometimes referred to as closed or "no-take" areas.** Some of the opposition to MPAs lies in resistance to "fencing the sea," reflecting a long tradition of open access. This opposition continues despite compelling empirical evidence and strong theoretical arguments indicating the value of using reserves as a tool to improve fisheries management, to preserve habitat and biodiversity, and to enhance the esthetic and recreational value of marine areas. The controversy persists because we lack a scientific consensus on the optimal design and use of reserves and we have only limited experience in determining the costs and benefits relative to more conventional management approaches. The current decline in the health of the ocean's living resources, an indication of the inadequacy of conventional approaches, and the increasing level of threat have made it more urgent to evaluate how MPAs and reserves can be employed in the United States to solve some of the pressing problems in marine management.

RECOGNIZING THE LIMITS

The ocean inspires awe; its vast expanse of water spans most of the earth's surface and fills the deep basins between continents. From the surface, the ocean appears uniform and limitless, seemingly too immense to feel the impacts of human activities. These perceptions led to the philosophy expressed by Hugo Grotius, a Dutchman in the 1600s, that the seas could not be harmed by human deeds and therefore needed no protection. His thinking established the principle of "freedom of the seas," a concept that continues to influence ocean policy despite clear evidence that human impacts such as overfishing, habitat destruction, drainage of wetlands, and pollution disrupt marine ecosystems and threaten the long-term productivity of the seas.

The flaw in the reasoning expressed by Grotius has been uncovered by research on the biology, chemistry, geology, and physics of the ocean. The sea is not a uniform, limitless expanse, but a patchwork of habitats and water masses occurring at scales that render them vulnerable to disturbance and depletion. The patchiness of the ocean is well known by fishers who do not cast their nets randomly but seek out areas where fish are abundant. There has been an increase in technology and fishing capacity that has led to a corresponding increase in the number of overfished stocks. Destruction of fish habitat as the result of dredging, wetland drainage, pollution, and ocean mining also contributes to the depletion of valuable marine species. As human populations continue to grow, so too does the pressure on all natural resources, making it not only more difficult, but also more critical to achieve sustainability in the use of living marine resources. These concerns have stimulated interest in and debate about the value and utility of approaches to marine resource management that provide more spatially defined methods for protecting vulnerable ocean habitats and conserving marine species, especially marine reserves and protected areas. **Based on evidence from existing marine area closures in both temperate and tropical regions, marine reserves and protected areas will be effective tools for addressing conservation needs as part of integrated coastal and marine area management.**

MANAGING MARINE RESOURCES

Management of living marine resources presents numerous challenges. The conventional approach typically involves management on a species-by-species basis with efforts focused on understanding population-level dynamics. For example, most fisheries target one or a few species; hence, managers and researchers have concentrated their efforts on understanding the population dynamics and effects of fishing on a species-by-species basis. Although this approach seems less complex, it does not resolve the difficulties of either managing multiple stocks or accurately assessing the status of marine species. This is compounded by the relative inaccessibility of many ocean habitats, the prohibitive expense of comprehensive surveys, and the complex dynamics and spatial heterogeneity of marine ecosystems. In addition, the species-specific approach may fail to address changes that affect productivity throughout the ecosystem. These changes may include natural fluctuations in ocean conditions (such as water temperature), nutrient over-enrichment from agricultural run-off and other types of pollution, habitat loss from coastal development and destructive fishing practices, bycatch of non-target species, and changes in the composition of biological communities after removal of either a predator or a prey species.

In addition to challenges presented by nature, management challenges arise from social, economic, and institutional structures. Regulatory agencies are charged with the difficult but important task of balancing the needs of current users with those of future users of the resource as well as the long-term interests of the general public. Regulatory actions intended to maintain productivity often affect the livelihoods of the users and the stability of coastal communities, generating pressure to continue unsustainable levels of resource use to avoid short-term economic dislocation. Finally, responsibility for regulating activities in marine areas, extending from estuarine watersheds to the deep ocean, is fragmented among a daunting number of local, state, federal, and international entities. This complexity in jurisdictional responsibility often places a major barrier to developing coordinated policies for managing ocean resources across political boundaries. Although the protected area concept, with its emphasis on management of spaces rather than species, is not new and has been used frequently on land, until recently there have been less support and few interagency efforts to institute protected areas as a major marine management measure. **MPA-based approaches will shift the focus from agency-specific problem management to interagency cooperation for implementing marine policies that recognize the spatial heterogeneity of marine habitats and the need to preserve the structure of marine ecosystems.**

To address these issues, the National Oceanic and Atmospheric Administration (NOAA), National Park Service, and Fish and Wildlife Service requested that the National Research Council's Ocean Studies Board assemble a committee of experts to examine the utility of marine reserves and protected areas for conserving marine resources, including fisheries, habitat, and biological diversity. Although there are other, equally important goals, for MPAs, including recreation, tourism, education, and scientific inquiry, examination of these objectives was not part of this committee's specified statement of task and hence receives less emphasis in this report. The committee was directed to compare the benefits and costs of MPAs to more conventional management tools, explore the feasibility of implementation, and assess the scientific basis and adequacy of techniques for designing marine reserves and protected areas. This report presents the findings of the study and provides recommendations for the application of marine reserves and protected areas as a tool in marine area management.

CONCLUSIONS AND RECOMMENDATIONS

MPA Design

Effective implementation of marine reserves and protected areas depends on participation by the community of stakeholders in developing the management plan. Federal and state agencies will need to provide resources, expertise, and coordination to integrate individual MPAs into the frameworks for coastal and marine resource management in order to meet goals established at the state, regional, national, or international level. **The lead agency will need to first identify all stakeholders, both on- and off-site, and then utilize methods of communication appropriate for various user groups.** Additionally, the needs and concerns of affected communities must be evaluated and considered when choosing sites for marine reserves and protected areas. Stakeholders should be encouraged to participate in the process by employing their expertise as well as considering their concerns.   Systematic social and economic studies will be required to recognize stakeholder groups, to assess the potential economic impacts of the MPA, and to determine community attitudes and goals.

The task of designing MPAs should follow four sequential steps: (1) evaluate conservation needs at both local and regional levels, (2) define the objectives and goals for establishing MPAs, (3) describe the key biological and oceanic features of the region, and (4) identify and choose site(s) that have the highest potential for implementation.

1. *Conservation Needs.* Local and regional conservation needs depend on the types of resources, the intensity and nature of human uses, and the physical and biological characteristics of the habitats. Consequently, the first step in planning an MPA is the identification and mapping of habitat types and living marine resources.

2. *Objectives and Goals.* The second step is the establishment of specific management goals for the proposed MPA. In most cases, the MPA will have multiple objectives such as protection of representative habitats, conservation of rare species, fish stock restoration or enhancement, or safeguarding of historical sites, among others. Ranking and prioritizing these objectives may be guided by local conservation needs and/or regional goals for establishing a network of MPAs. Conflicting objectives may require negotiation, trade-offs, and consideration of social and economic impacts.

There are multiple goals for establishing MPAs, such as conserving biodiversity, improving fishery management, protecting ecosystem integrity, preserving cultural heritage, providing educational and recreational opportunities, and establishing sites for scientific research. However, the focus of this report is on conserving biodiversity and improving fishery management through the use of

MPAs and marine reserves. To promote *biodiversity*, the siting criteria for an MPA or reserve may include habitat representation and heterogeneity, species diversity, biogeographic representation, presence of vulnerable habitats or threatened species, and ecosystem functioning. To improve *fishery management*, site choice may depend on the locale of stocks that are overfished to provide insurance against stock collapse or to protect spawning and nursery habitat. Alternatively, a site may be selected to reduce bycatch of nontarget species or juveniles of exploited species.

3. *Biological and Oceanic Features.* Evaluating the suitability of potential sites under these criteria requires the collection and integration of information on the life histories of exploited or threatened species (e.g., location of spawning and nursery sites, dispersal patterns) and the oceanic features of the region. The latter may include water current and circulation patterns, identification of upwelling zones and other features associated with enhanced productivity, water quality (nutrient inputs, pollution, sedimentation, harmful algal blooms), and habitat maps.

4. *Site Identification.* Distilling the desired properties of an MPA into a zoning plan that specifies size and location of reserves requires matching the biological and oceanic properties to meet the specified objectives. Guidelines and general principles that can be applied to this task are described below.

Identifying Locations

Choice of sites for MPAs should be integrated into an overall plan for marine area management that optimizes the level of protection afforded to the marine ecosystem as a whole because the success of MPAs depends on the quality of management in the surrounding waters. In coastal areas specifically, MPAs will be most effective if sites are chosen in the broader context of coastal zone management, with MPAs serving as critical components of an overall conservation strategy. Management should emphasize spatially oriented conservation strategies that consider the heterogeneous distribution of resources and habitats. This may include selecting MPA sites based on the location of terrestrial protected areas. For example, locating an MPA adjacent to a national park may provide complementary protections for water quality, restoration of nursery habitat, and recovery of exploited species. Often a single MPA will be insufficient to meet the multiple needs of a region and it will be necessary to establish a network of MPAs and reserves, an array of sites chosen for their complementarity and ability to support each other based on connectivity. Connectivity refers to the capacity for one site to "seed" another location through the dispersal of either adults or larvae to ensure the persistence and maintenance of genetic diversity for the resident protected species.

Sites that meet the ecological and oceanographic criteria must also be evaluated with respect to the patterns of stakeholder use in those areas. Site identification should maximize potential benefits, minimize socioeconomic conflicts to the extent practicable, and exclude areas where pollution or commercial development have caused problems so severe that they would override any protective benefit from the reserve and so intractable that the situation is unlikely to improve.

Determining Size

The optimal size of marine reserves and protected areas should be determined for each location by evaluating the conservation needs and goals, quality and amount of critical habitat, levels of resource use, efficacy of other management tools, and characteristics of the species or biological communities requiring protection. The boundaries of many MPAs, such as those in the National Marine Sanctuary Program, have been drawn based on specific topographic features, but deciding on the size of marine reserves (i.e., no-take zones) requires greater consideration of the biological features to meet specific management goals. In many cases, specific attributes of the locale (saltmarsh habitat, spawning and nursery grounds, special features such as coral reefs, seamounts, or hydrothermal vents) will determine the size of an effective reserve. In other cases, the dispersal patterns of species targeted for protection, as well as the level of exploitation, should be considered in deciding how much area to enclose within a reserve. Achieving the various marine management goals outlined in this report will require establishing reserves in a much greater fraction of U.S. territorial waters than the current level of less than 1%. Proposals to designate 20% of the ocean as marine reserves have focused debate on how much closed area will be needed to conserve living marine resources. The 20% figure was originally derived, in part, from the value fishery managers once recommended for conservation of a fish stock's reproductive potential (i.e., the target spawning potential ratio). For sedentary species, protecting 20% of the population in reserves will help conserve the stock's reproductive capacity and may roughly correlate with 20% of that species' habitat. However, the optimal amount of reserve area required to meet a given management goal may be higher or lower depending on the characteristics of the location and its resident species, as described in Chapter 6 and summarized in Table 6.3 of this report. Size optimization generally will require adjustments to the original management plan based on reserve performance, as determined through research and monitoring. Hence, the first priority for implementing reserve sites should be to include valuable and vulnerable areas rather than to achieve a percentage goal for any given region.

Designating Zones and Designing Networks

Zoning should be used as a mechanism for designating sites within an MPA to provide the level of protection appropriate for each management

goal. In many instances, multiple management goals will be included in an MPA plan and zoning can be used to accomplish some of these goals. These zones may include "ecological reserves" to protect biodiversity and provide undisturbed areas for research, "fishery reserves" to restore and protect fish stocks, and "habitat restoration areas" to facilitate recovery of damaged seabeds. Frequently, an MPA is established initially to protect a site from threats associated with large-scale activities such as gravel mining, oil drilling, and dredge spoil disposal. Under these MPA-wide restrictions, there is an opportunity to resolve other conflicting uses of marine resources through zoning of areas within the MPA. Networking to provide connectivity (see section "Identifying Locations") should be considered in both zoning and siting of MPAs to ensure long-term stability of the resident populations.

Monitoring and Research Needs

Monitoring

The performance of marine reserves should be evaluated through regular monitoring and periodic assessments to measure progress toward management goals and to facilitate refinements in the design and implementation of reserves. Marine reserves should be planned such that boundaries and regulations can be adapted to improve performance and meet changes in management goals. There are three tasks that should be included in a well-designed monitoring program: (1) assess management effectiveness; (2) measure long-term trends in ecosystem properties; and (3) evaluate economic impacts, community attitudes and involvement, and compliance.

Monitoring programs should track ecological and socioeconomic indicators for inputs to and outputs from the reserve at regular time intervals. Inputs might include water quality, sedimentation, immigration of adults and larvae of key species, number of visitors, and volunteer activities. Outputs might include emigration of adults and larvae of key species, changes in economic activity, and educational programs and materials. Within the reserve, monitoring efforts should assess habitat recovery and changes in species composition and abundance.

Research

Research in marine reserves is required to further our understanding of how closed areas can be most effectively used in fisheries and marine resource management. Reserves present unique opportunities for research on the structure, functioning, and variability of marine ecosystems that will provide valuable information for improving the management of marine resources. Whenever possible, management actions should be planned to facilitate rigorous examination of the hypotheses concerning marine reserve design and implementation. Research in reserves could provide estimates for important parameters in fishery models such as natural mortality rates and dispersal properties of larval, juvenile, and adult fish. Other research programs could test marine reserve design principles such as connectivity or the effect of reserve size on recovery of exploited species. **Modeling studies are needed both to generate hypotheses and to analyze outcomes for different reserve designs and applications.**

Institutional Structures

Integration of management across the array of federal and state agencies will be needed to develop a national system of MPAs that effectively and efficiently conserves marine resources and provides equitable representation for the diversity of groups with interests in the sea. The recent executive order issued by the White House on May 26, 2000, initiates this process through its directive to NOAA (Department of Commerce) to establish a Marine Protected Area Center in cooperation with the Department of the Interior. The goal of the MPA Center shall be "to develop a framework for a national system of MPAs, and to provide Federal, State, territorial, tribal, and local governments with the information, technologies, and strategies to support the system." Establishment of a national system of MPAs presents an opportunity

- to improve regional coordination among marine management agencies;
- to develop an inventory of existing MPA sites; and
- to ensure adequate regulatory authority and funds for enforcement, research, and monitoring.

Effective enforcement of MPAs will be necessary to obtain cooperation from affected user groups and to realize the potential economic and ecological benefits. Also, coordination among agencies with different jurisdictions will improve the representation of on-site and off-site user groups so that the general public's cultural and conservation values, as well as commercial and recreational activities, receive consideration. Under current management approaches, these interests are often addressed by different agencies independently of each other and may result in short-term policies that are inconsistent with the nation's long-term goals.

Conclusion

What are the consequences of not developing a national system of marine reserves and protected areas? Are conventional management strategies sufficient to ensure that our descendents will enjoy the benefits of the diversity and abundance of ocean life? One purpose of this report is to compare conventional management of marine resources with proposals to augment these management strategies with a system of protected areas. Although it may seem less disruptive to rely on the familiar, conventional management tools, there are costs associated with maintaining a status quo that does not meet conservation goals. Hence, our relative inexperience in using marine reserves to manage living resources should not serve as an argument against their use. Rather, it argues that implementation of reserves should be incremental and adaptive, through the design of areas that will not only conserve marine resources, but also will help us learn how to manage marine species more effectively. The dual realities that the earth's resources are limited and that demands made on marine resources are increasing, will require some compromise among users to secure greater benefits for the community as a whole. Properly designed and managed marine reserves and protected areas offer the potential for minimizing short-term sacrifice by current users of the sea and maximizing the long-term health and productivity of the marine environment.

1

Introduction

There is broad recognition that the oceans and their living resources are under stress. Increasing use by humans, especially in the coastal zone but increasingly offshore as well, have damaged marine habitats and led to overfishing of many marine fish stocks. Significant numbers of marine organisms, including mammals, birds, and turtles, as well as some commercially harvested fish and shellfish, are now threatened or endangered. The threats of further habitat damage, loss of species, and loss of genetic diversity—all attributable to human actions—in addition to increasing problems from overfishing, loom imposingly on the horizon. Clearly, new management approaches or options must be considered to stem the damage and ensure that marine ecosystems and their unique features are protected and restored. In this regard, marine reserves and protected areas are more often proposed as major tools to relieve stress on marine resources and ecosystems. This report evaluates the use of protected areas and reserves for the conservation of living marine resources, and makes recommendations on their potential implementation as a management tool in marine waters of the United States.[1]

The oceans occupy more than 70% of the earth's surface and 95% of the biosphere and once were thought to be so vast that it was judged inconceivable that human activities might significantly alter the structure and functioning of

---

[1] Marine waters in the United States refers to the exclusive economic zone of the coastal states and territories.

marine ecosystems. However, it is now obvious that the seas feel the stamp of heavy human use from industries such as fishing and transportation, the effects of waste disposal, excess nutrients from agricultural runoff, and the introduction of exotic species. The cumulative effect on marine ecosystems has attracted public attention and enhanced public concern for ocean resources, unique habitats, and the threats to continuing marine ecosystem productivity.

Most of the world's fish stocks are now heavily exploited. As many as 25 to 30% are overfished, and another 44% are fully exploited (Garcia and Newton, 1997; FAO, 1999; NRC, 1999a). In Europe, the impact of fishing on fish population abundance became evident when naval activities and extensive minefields closed the North Sea fishery during World Wars I and II. While catches prior to the wars were declining, there were dramatic recoveries immediately afterwards when it was safe to resume fishing activity (Gulland, 1974; Cushing, 1975). These recoveries supported the idea that time and area closures could be established to restore and protect overfished stocks.

Given the growing perception that current management of marine resources and habitats is insufficient, interest is growing in approaches to ensure the continuing viability of marine ecosystems. Over the past century, concern about the rapid loss of wilderness lands led to establishment of protected areas, reserves, and parks in terrestrial ecosystems where human activities are much restricted or at least curtailed. Generally, the objective in these areas is to protect or restore ecosystems, to preserve the natural beauty of the landscape, and to support the survival of native species. The public accepts these concepts and cherishes protected areas such as national parks and wildlife refuges. Yet this approach has not transferred to the marine environment. The effectiveness of marine reserves and marine protected areas (MPAs) is debated passionately by advocates and detractors, even though more than a thousand MPAs have been established around the globe. Similar to terrestrial protected areas, advocates promote their benefits as insurance against overexploitation, conservation of biodiversity, and protection of habitat. Their potential as tools for fisheries management is recognized by many scientists (Bohnsack, 1998). However, few MPAs have been evaluated critically to determine to what extent they benefit exploited species.

There have been numerous attempts to develop terms and definitions to encompass the array of applications of MPAs in marine conservation. In principle, the committee accepts the classification scheme developed by the International Union for the Conservation of Nature and Natural Resources (IUCN, see Appendix F) which applies to both terrestrial and marine protected areas (IUCN, 1994). The six categories in this scheme provide a mechanism for assessing the status of protected areas internationally. However, the specificity provided by the IUCN classification makes it impractical for quick reference to the more general goals of MPAs described in this report. Therefore, the committee defined a simplified list of terms for the various types of protected areas, listed here in order of increasing levels of protection:

• *Marine Protected Area*—a discrete geographic area that has been designated to enhance the conservation of marine and coastal resources and is managed by an integrated plan that includes MPA-wide restrictions on some activities such as oil and gas extraction and higher levels of protection on delimited zones, designated as fishery and ecological reserves within the MPA (see below). Examples include the Florida Keys National Marine Sanctuary and marine areas in the National Park system, such as Glacier Bay.

• *Marine Reserve*—a zone in which some or all of the biological resources are protected from removal or disturbance. This includes reserves established to protect threatened or endangered species and the more specific categories of fishery and ecological reserves described below.

• *Fishery Reserve*—a zone that precludes fishing activity on some or all species to protect critical habitat, rebuild stocks (long-term, but not necessarily permanent, closure), provide insurance against overfishing, or enhance fishery yield. Examples include Closed Areas I and II on Georges Bank, implemented to protect groundfish.

• *Ecological Reserve*—a zone that protects all living marine resources through prohibitions on fishing and the removal or disturbance of any living or non-living marine resource, except as necessary for monitoring or research to evaluate reserve effectiveness. Access and recreational activities may be restricted to prevent damage to the resources. Other terms that have been used to describe this type of reserve include "no-take" zones and fully-protected areas. The Western Sambos Reserve in the Florida Keys National Marine Sanctuary provides an example of this type of zoning.

Defining the goals and objectives from among the myriad that may exist is a prerequisite for determining the appropriate level of protection for an MPA (Agardy, 1997; Allison et al., 1998). The objectives must be clear with respect to expectations of performance and the degree to which human activities, including extractive uses and tourism, must be restricted to achieve goals. Promoting fishery management goals and objectives may require different criteria for designating and implementing MPAs, than for protecting unique habitats or biological diversity.

Decisions regarding location, size, and linkages between MPAs and other components of ecosystems must be considered. Adopting MPAs as a major management tool will require a shift in management emphasis from single-species management to spatial management. Oceanographic features, bathymetry, hydrography, and the transport of organisms into or out of MPAs can be critical factors in MPA design. The human element, including stakeholder involvement in the planning and implementation stages for MPAs, is critical in determining whether an MPA will successfully meet its objectives or whether it will result in resentment and noncompliance by individuals and communities that face restrictions on current and future uses.

Although MPAs currently occupy less than 1% of the marine environment, their use is increasing throughout the world (Kelleher, 1999). Recent recognition that fishing activities, especially bottom trawling, but also dredging, fish traps, and longlines, can alter or destroy habitat and that many fisheries in the United States and globally are overfished (Dayton et al., 1995; NOAA, 1996b) demonstrates the need to explore alternative approaches for protecting and managing the sea. Many studies are now under way to evaluate the potential of fishery reserves as a complementary or alternative approach to conventional fishery management and to determine if reserves can successfully conserve fish stocks, while preserving biodiversity and protecting habitat. Degradation of marine ecosystems also results from coastal land use and watershed problems. Establishment of MPAs and reserves can prompt improved management of land-based activities that impact estuarine and marine habitats. Advocates argue that only reserves can provide insurance against management failures resulting from insufficient research or uncertainty intrinsic to complex and poorly understood marine ecosystems. This argument has been challenged by others who view conventional management approaches, if rigorously applied, as both effective and less disruptive to resource users. In this sense, it is important to distinguish between the different objectives of marine reserves, some focusing on issues of biological diversity and others directed at managing fisheries, when evaluating them as management tools. Highlights of that debate are captured in this report.

WHY MPAs?

As management becomes more integrated and holistic, MPAs will take on greater importance as a tool for conserving marine resources. In particular, MPAs have been proposed as an integral component of marine and coastal zone management, with establishment of regional networks of MPAs as a means to improve overall governance of the coastal ocean (Done and Reichelt, 1998). However, implementation has been hindered by a lack of consensus on how to design MPAs to maximize their utility. The extent of current threats to marine resources may justify establishment of MPAs and reserves, despite the lack of experience, using an adaptive management approach to modify the design as knowledge and experience increase.

Declines in biological diversity and productivity can be precipitated in many cases by fishing and other human interventions (e.g., dams, dredging, coastal development, and wetland losses, introduced species, tourism and recreational activities). These declines have spurred efforts to institute alternative management approaches that will conserve and, where needed, restore biological diversity and productivity. MPAs, like their counterparts in terrestrial ecosystems, can be used to protect critical or threatened habitats in order to foster restoration of biological communities and their productivities. Importantly, establishment of MPAs may motivate communities to increase their stewardship of the ocean through stricter land use policies and pollution controls.

Many observers believe that conventional management has not supported sustainable marine fisheries (Ludwig et al., 1993). Further, scientists have found that habitats on the seabed, along with the diverse communities of organisms that they support, are being degraded by fishing and other human activities (Watling and Norse, 1998; Langton and Auster, 1999). In response, there are demands for new resolve in the form of precautionary management and adoption of ecosystem approaches to fisheries management (NMFS, 1999; NRC, 1999a). The challenges are to prevent overfishing, protect marine habitats, and restore biodiversity. In the United States, the Magnuson-Stevens Fishery Conservation and Management Act (NOAA, 1996a) requires the elimination of overfishing and protection of essential habitats. Marine reserves are proposed as one tool that can provide insurance against uncertainties in fisheries science and promote the conservation and restoration of fish habitats.

Users of marine resources do not always embrace the concept of MPAs or welcome them when instituted. Stakeholders may distrust managers and scientists, especially when confronted with the possibility of losing their customary access privileges. Also, competing users (e.g., commercial fishers, recreational fishers, divers, farmers, developers, realtors, industrial concerns) may perceive inequities in the allocation of privileges in MPAs. These problems are especially prevalent when stakeholders are not fully involved in the design and planning of MPAs (Kelleher and Recchia, 1998) and often lead to opposition and hostility.

This report reviews the state of knowledge of marine reserves and protected areas and evaluates their utility for promoting and conserving biodiversity, improving fishery management, and protecting habitats in the sea. The scope of the committee's task was broad. With respect to fisheries, it included a comparison of reserves with conventional "command-and-control" fisheries management (regulating catch and fishing effort) and also with emerging "rights-based" approaches, such as individual fishing quotas (IFQs) (NRC, 1999b). Reserves also were evaluated with respect to societal needs and concerns. The potential for MPAs and reserves to affect both direct and indirect users of marine resources was recognized, and the need for adaptive responses by managers with respect to design was noted. As defined in the Statement of Task below, the focus of this study was on conservation of living marine resources; hence, other potential goals of MPAs such as protection of cultural artifacts, increased educational opportunities, and enhancement of tourism, although mentioned, are not examined in detail.

STATEMENT OF TASK

The prospectus for this study defines four tasks for the committee as follows:

1. examine the utility of marine reserves and protected areas to conserve marine biological diversity and living resources, including fisheries;
2. compare benefits and costs of this approach to more conventional tools;
3. explore the feasibility of implementing marine reserves and protected areas; and
4. assess the scientific basis and adequacy of techniques used for the location, design, and implementation of marine reserves and protected areas, including their successes for management of fisheries.

The project reviews the design, implementation, and evaluation of marine reserves and protected areas, using examples from the United States as well as Australia, New Zealand, Canada, and other countries in which they have been implemented. The adequacy of current efforts to use marine protected areas and reserves is assessed both as a management approach for restoring declining fish stocks and as a tool for conserving marine biological diversity. This report recommends ways to improve the implementation of marine protected areas and reserves, and identifies future research that could assist in implementing these tools more effectively.

STUDY APPROACH AND REPORT ORGANIZATION

This study evolved from a confluence of interests in the timely and controversial topic of setting aside areas in the ocean for the conservation and preservation of living marine resources. Primary funding was supplied by the National Oceanic and Atmospheric Administration's National Marine Fisheries Service and National Marine Sanctuaries Program, with additional funds from the Department of the Interior's Fish and Wildlife and National Park Services. The committee held four information-gathering meetings at the following sites: Washington, D.C.; Islamorada, Florida; Monterey, California; and Seattle, Washington. Speakers from each region were invited to address the committee and time was allowed for public comments (Appendix D).

In organizing this report, the committee sought to cover the more difficult issues surrounding the design and implementation of marine reserves and protected areas. Chapter 2 describes the differences between marine and terrestrial ecosystems that influence both the goals and the design of protected areas. Specific goals for establishing protected areas in marine environments are also described in that chapter. Because much of the interest in reserves and MPAs has emerged from the perceived failure of conventional fisheries management strategies, the strengths and weaknesses of these conventional approaches are explored in Chapter 3. Chapter 4 describes the values, expected costs and benefits, and need for stakeholder involvement in identifying goals and establishing management plans for MPAs and reserves. Chapter 5 presents both the theoretical arguments and the empirical evidence for marine reserves in the form of a literature review. Planning and design are critical steps for successful establishment of MPAs and reserves, and these issues are presented in Chapter 6. After a marine reserve has been established, monitoring and research are needed to evaluate the effectiveness of the reserve in attaining management goals. Chapter 7 describes approaches that can be used to evaluate reserve performance. Chapter 8 describes the international history of MPAs and critiques the current system of MPAs and reserves in the United States. Finally, in Chapter 9, the committee presents its conclusions and recommendations.

2

Conservation Goals

CONSERVATION GOALS ON LAND AND IN THE SEA

Terrestrial reserves and protected areas have a long history compared to marine protected areas (MPAs) and many lessons can be learned for application to MPAs. Although MPAs will require different design features than terrestrial protected areas, the motivations for creating them are similar and include maintaining essential ecological processes, preserving biological diversity, ensuring the sustainable use of species and ecosystems, and protecting cultural heritage sites.

Differences in approaches to the conservation of marine and terrestrial areas reflect both (1) differences in ecosystem processes and (2) differences in historical perceptions and regulatory frameworks.

Differences Between Marine and Terrestrial Ecosystems

Much of the theory of conservation biology has focused on developing management strategies to protect terrestrial wildlife. However, application of these theories to marine conservation has been debated. The discussion that follows highlights some of the differences that may affect application of terrestrial-based models to conserve marine species.

Marine and terrestrial ecosystems differ in that marine ecosystems are relatively open, while terrestrial ecosystems have more discrete boundaries. As a consequence, migration and dispersal of organisms in various life stages are more characteristic of marine ecosystems. Other dissimilarities originate from differences in spatial scales of the habitats and contrasts between life strategies in water and on land. Marine ecosystems also may be more variable than terrestrial ecosystems, especially on shorter time scales. Marine ecosystems are subject to the physics of the surrounding medium and respond to forces such as tides, circulation patterns, and decadal shifts in overall productivity, whereas terrestrial ecosystems are more internally controlled by the life processes of the dominant organisms (e.g., trees) and may change only slowly, sometimes on century time scales, unless humans intervene (Steele, 1985, 1991, 1996).

On land, survival of rare or endangered species is especially dependent on habitat, which often plays a decisive role in identifying areas worthy of protection. The case for protection of a terrestrial area to save a species from extinction has provided powerful arguments for garnering public support. Habitat destruction accounts for about 36% of animal extinctions whose cause is known (compared to 23% due to hunting and 39% due to introduced species) and is thought to be even greater for the extinction of terrestrial species where the cause is unknown (Groombridge, 1992). As people increase their use of the land, habitats to support terrestrial species will continue to decline, both from destruction and from fragmentation into areas too small to support indigenous populations.

Human populations appear to have less impact on marine habitats because people do not live in the ocean and thus are less aware of the change. The loss of marine habitat, except for wetlands and estuarine marshes, has been documented infrequently, and population declines or extinctions in marine species are more often attributed to overexploitation. Historically, the concept of conserving critical habitat for endangered marine species has been applied mostly to marine mammals, sea turtles, and sea birds, with only occasional application to endemic fishes or invertebrates (Kelleher and Kenchington, 1992). However, the dramatic loss of coastal wetlands (NRC, 1992) and recent descriptions of the impacts of trawling gear on the seabed (Watling and Norse, 1998), among other stresses, have led to increased attention to the vulnerability of some marine species to extinction from loss of habitat (Roberts and Hawkins, 1999). Even more common is the decrease in genetic diversity from the loss of distinct populations associated with habitat at a discrete site.

In selecting areas for protection, several concepts applied to terrestrial reserves are also important for marine reserves, including sources and sinks, dispersal range, and metapopulations (see Chapter 6). When the range of a species is large and the density of the population is relatively low, it may be impractical to design a reserve that is large enough to protect the species. On land, the solution may require establishing several reserves connected by corridors that allow the physical passage of species. In the ocean, water provides the corridor, and the design issue rests on an understanding of currents and circulation patterns or other oceanographic features that will either facilitate or impede the dispersal of individuals among reserves (see Chapter 6). Also, even sedentary and nonmigratory marine species commonly have a mechanism for dispersal through a reproductive larval stage that provides a level of insurance against their localized extinctions. As a consequence of these broad dispersal ranges, many marine species do not show genetic isolation even over large distances (Palumbi, 1992; also see Chapter 5).

Although few marine organisms are known to face extinction as a consequence of endemism and threatened habitat, there are important exceptions. The American Fisheries Society (AFS) recently recognized species vulnerable to extinction. These species generally are long-lived, mature slowly, have low fecundity, are closely associated with particular habitats, and are exceptionally vulnerable to fishing or other anthropogenic stresses. High-seas predators (e.g., tunas, marlins, swordfish, sharks), although not closely associated with seabed habitats, also are vulnerable. In a historic move, AFS has adopted policies that acknowledge the special needs of such species, which may become threatened or endangered if not managed wisely. AFS has recommended MPAs as one management tool to protect species at risk of extinction (Musick, 1999: Coleman et al., 2000).

In the marine environment, mobile species such as fish, marine mammals, and sea turtles, move in three dimensions and have a much greater ability to migrate over long distances than is common for organisms in terrestrial ecosystems. This makes it more difficult to identify discrete populations and blurs the apparent boundaries of marine ecosystems. Also, the relative openness and fluidity of marine ecosystem boundaries increase the likelihood that they will be subject to external influences such as pollution from surrounding lands and waters (Steele, 1985, 1991).

Another difference between terrestrial and marine ecosystems is that most seafood is obtained by fishing, not farming. Wild stocks of fish, not aquaculture, remain the major source of the world's seafood (New, 1997; Naylor et al., 1998, 2000), while land-based agriculture, not hunting, is the main terrestrial food source. Therefore, the continued supply of seafood for human consumption is dependent on sustainable fishing practices for the foreseeable future or until mariculture becomes independent of fish-based food sources. Finally, in contrast to the plants and herbivores that dominate terrestrial food production, most exploited fish species are carnivores, and their depletion may have cascading influences on marine food webs, such as the expansion of herbivore populations and subsequent declines in algal coverage from increased grazing pressure.

Differences in Human Perceptions and Use of Marine and Terrestrial Areas

In socioeconomic terms, a fundamental difference between the use and management of resources in the sea and on land arises from historical perceptions or definitions of ownership and the laws and conventions that govern these activities. On land, problems arising from common property rights have been summarized as "the tragedy of the commons" (Hardin, 1968, 1998). The failure of communities to limit use of the commons by individuals in the cause of overall community interest and sustainability has led to a shift in most countries to private or government ownership of most land areas. This shift imbues property owners with a strong incentive to protect the land and its resources from overuse and destructive activities, thus empowering the owners to act as stewards of the land. In contrast, coastal waters have been considered part of the public trust in the United States, a concept applied since colonial times based on English common law, with origins extending as far back as Roman times (Hanna et al., 2000). Internationally, only recently have nations acted to establish ownership of the seabed and overlying waters through declaration of territorial seas and exclusive economic zones (EEZs). These levels of ownership are far more limited than standards applied to most land areas. Nevertheless, since the 1970s there has been a notable shift toward granting privileged access to marine resources for some groups while excluding others. International conventions regarding jurisdiction over marine waters are discussed in Chapter 8.

Outside of EEZs, the concept of ownership of portions of the sea or seabed is slowly increasing, as expressed principally in the United Nations Convention on the Law of the Sea (UNCLOS). Some maritime nations, including the United States, are party to neither UNCLOS nor the Convention on Biological Diversity. Consequently, few areas outside territorial waters are fully regulated with respect to international use. For example, the only marine areas outside national territorial waters in which ship activities are restricted by international agreement are part of the Great Barrier Reef Marine Park and the Sabana-Camaguey Archipelago off the coast of Cuba. These areas were declared to be "particularly sensitive sea areas" by resolutions of the International Maritime Organization (IMO) in 1990 and 1997 respectively, under the provisions of the International Convention for the Prevention of Pollution from Ships (MARPOL).

Implications for MPAs

The general public, as well as special interest groups, cherishes the right to use marine areas and resources without restriction. Historically, attempts by government to limit this freedom, even for the benefit of users, have been fought bitterly by those users. For instance, the National Marine Sanctuary Program has struggled to gain public acceptance of fishing restrictions or prohibitions within areas designated as ecological reserves. In the Florida Keys National Marine Sanctuary, less than 0.5% of the sanctuary is closed to all fishing, and most of the other national marine sanctuaries have no areas closed to fisheries. It is difficult to change the perception that access to marine resources is a right because the open-access doctrine has deep roots in the United States.

GOALS OF MARINE RESERVES AND PROTECTED AREAS

To analyze the usefulness of MPAs and reserves as tools for environmental management, it is important to recognize that this approach has been proposed to meet a wide variety of goals. Typically, MPAs will be established to meet multiple goals, enhancing the efficiency and optimizing the value of the area in the context of coastal and marine area management. These goals are classified into the six categories discussed below.

Conservation of Biodiversity and Habitat

Calls for the preservation of biodiversity and natural habitats stem from many different concerns, ranging from the aesthetic to the economic. A strong component of human nature involves an appreciation of, and a desire to understand, the world around us. People recognize the value of continuity with the past and into the future, and there is a strong desire to perpetuate representative habitats for future generations. A manifestation of this is the fact that many human cultures have established and protected parks, sometimes for thousands of years. This is the heritage value of representative marine habitats and ecosystems. Marine reserves offer an important if not unique means of protecting marine wilderness for the future use of humanity.

Preservation of biodiversity and habitat also has contemporary value because of the ecosystem services provided by natural marine communities. Those communities are threatened by habitat loss and depletion of economically valuable species (Murray and Ferguson, 1998; NRC, 1999a). Examples of marine ecosystem services include goods (e.g., seafood, shells, aquarium fish), life support processes (e.g., carbon sequestration, nutrient recycling), quality of life (beauty, enjoyment of natural seascapes), and potential future uses (drug discovery, genetic diversity) (Daily et al., 2000). Marine reserves function in several ways to conserve biodiversity and habitat, two goals that are inextricably linked.

Protect Depleted, Threatened, Rare, or Endangered Species or Populations

Although documented cases of marine species at risk of extinction are rare, this may reflect the lack of research rather than actual low incidence (Roberts and Hawkins, 1999). Many local marine populations have indeed been severely depleted or are functionally extinct (Dayton et al., 1998), with a potential loss of genetic diversity. For example, giant clams (*Tridacna gigas*) have been extirpated from several island archipelagoes in the Pacific Ocean by overfishing (Wells, 1997); sawfish (*Pristis pectinata*) have been eliminated from many estuaries on the east coast of the United States by fishing (Poss, 1998); the white abalone (*Haliotis sorenseni*) has recently been declared a candidate for the federal endangered species list and may become the first marine invertebrate known to be fished to extinction (Tegner et al., 1996); and the totoaba (*Totoaba macdonaldi*), once so abundant in the Gulf of California that millions were landed and their bodies used as fertilizer, now hovers on the brink of global extinction as a consequence of overfishing, loss of estuarine spawning habitat (due to diversion of water from the Colorado River), and bycatch of juveniles in shrimp trawls (Cisneros-Mata et al., 1997). Reserves may be established with the specific goal of protecting such species or preserving habitat considered critical for their survival.

Preserve or Restore the Viability of Representative Habitats and Ecosystems

By preserving representative ecosystems, marine reserves are likely to ensure the conservation of diverse species assemblages and maintain genetic diversity. Although the greater openness of marine systems and the dispersal capabilities of marine organisms help reduce the likelihood of extinction through habitat loss, maintaining the full range of habitat types is necessary for food and shelter to support different stages in the life histories of these organisms and to support ecological processes such as nutrient recycling.

Some habitats are heavily impacted by bottom trawling, pollution, dredging, and oil and gas drilling. Distinctive habitats can be critical to many types of species, for example, as spawning aggregation sites or as juvenile nurseries. These habitats may range from coral reefs to seamounts to mangroves to kelp forests. Losses in biodiversity through habitat destruction generally are unintended (which is *not* to say unforeseen) consequences of capturing one or more target species using technology that massively impacts habitat and nontarget organisms (Dayton et al., 1995). This point is brought home most forcefully perhaps by considering benthic habitats in which trawling activities have led to massive destruction of physical and biological features and, as a consequence of this destruction, profound alteration of ecosystem structure and function (Thrush et al., 1998). In the case of pelagic fishing, bycatch is likely to be the key negative side effect on nontarget species, but in the case of benthic trawling, the entire ecosystem faces massive disturbance (Watling and Norse, 1998). When essential or significant habitats can be identified, they can be protected by the implementation of reserves. Marine reserves can also be established to help restore disturbed critical habitat.

Fishery Management

Fishery reserves can improve fishery management in various ways, depending on the characteristics of the resources, their fisheries, and the management system in place. The following goals of reserves related to fishery management are identified here, with the understanding that such fishing closures are likely to be embedded in larger management areas subject to different types of fishing and environmental regulations.

Control Exploitation Rates

Reserves can help control or reduce exploitation rates mainly in two ways. First, for species of low adult mobility, reserves can be an effective tool to control catch rates by directly protecting some fraction of the population from the effects of fishing. Indeed, much of the impetus for establishing reserves has come from experience with sedentary reef species, which have been severely overfished in the past, and where fishing pressure has proved difficult to control by other means. In these cases, fishery reserves may help enhance depleted fish stocks, provided the hotspots of reproduction created within the reserves are large and replenish the populations outside reserve boundaries.

A second way in which reserves can reduce fishing rate is by diverting fishing effort away from areas of high fish density areas where fish are less vulnerable. This can be effective in fisheries that are managed by limiting the total amount of fishing effort or in fisheries that are essentially unregulated. The large closed areas now in place on Georges Bank, for example, have been found to contribute significantly to reducing fishing mortalities of cod (*Gadus morhua*) and yellowtail flounder (*Limanda ferruginea*), fisheries managed by limiting days at sea. The rebuilding plan for these depleted stocks reduced the catch both by reducing days at sea (i.e., placing tighter effort controls) and by reducing the efficiency of the fishing effort through the implementation of large closed areas on preferred fishing grounds. These closures displaced effort to areas with lower fish densities, thereby lowering the catch per day fished (Murawski et al., 2000). The rebuilding plan for these depleted stocks hence reduced the catch both by reducing days at sea and by reducing the efficiency of the fishing effort. A potential drawback of this approach is that lowering fishing efficiency may spread the impacts of fishing (bycatch and habitat alteration) over a larger area.

When conventional means of regulating fishing such as catch quotas or effort limitations are not an option (because they are either impractical, unenforceable, or too costly, or because the information required is simply not available), large spatial closures placed on areas of high fish concentrations could become the primary regulatory tool. Conventional, single-species management tools, for example, rapidly become impractical in multispecies fisheries when the fleet cannot selectively target individual stocks. Effort cannot be fine-tuned to meet individual species targets. Implementation of catch quotas by species leads to complex arrays of limits on the catch by species per fishing trip, which not only result in high levels of discard but also may fail to reduce fishing mortality. Reserves may be the only practical way to protect the most vulnerable species in these complexes or stocks that have been overfished in the past. Even if no directed fisheries were allowed on these overfished species, rebuilding may be possible only if areas in which significant incidental catch occurs are closed. This may be the situation for bocaccio (*Sebastes paucispinis*) on the west coast of the United States; a rockfish for which a rebuilding plan has recently been approved (www.pcouncil.org/Groundfish).

Protect Critical Stages of the Species' Life History

Protecting nursery grounds, or areas where discards of juvenile fish would be high if they were open to fishing, has been one of the most common reasons for establishing reserves in the past. Closing nursery grounds can be very effective for stock conservation because for most exploited species, a reduction in the mortality of juveniles has a larger payoff in terms of increased mature biomass than a proportional reduction in adult mortality (Horwood et al., 1998). Where the habitat of the nursery grounds is itself vulnerable to damage from fishing, it is clear that reserves will help to protect both the resident juveniles and the ecosystem on which their survival and maturation depend.

Another important goal of reserves has been to protect areas where fish aggregate to spawn. Beyond the possible reduction of fishing efficiency, as discussed above, protecting spawning aggregations may be important (1) for species that exhibit complex reproductive behaviors that would be disrupted by fishing operations; (2) when survival of eggs, larvae or juveniles present on the spawning grounds would be imperiled if fishing were permitted; or (3) when fish aggregate in such high densities to spawn that controlling catches on the aggregations would be difficult (Johannes, 1998; also see Box 2-1).

Reduce Secondary Fishing Impacts

The effects of fishing go well beyond the capture of the target species (Dayton et al., 1995; Watling and Norse, 1998). As noted earlier, the habitat on which targeted species depend may be severely affected by fishing. Depending on the gear used, fisheries may profoundly alter the characteristics of the bottom and benthic ecosystems (Goñi, 1998; Watling and Norse, 1998; Hall, 1999). The protection of benthic habitats from destruction not only will maintain biodiversity, but also may enhance the fishery in question if the target species, at some stage(s) of its life history, depends on critical habitat or components of the ecosystem perturbed by fishing. Many economically valuable species have larval or juvenile stages that depend on particular substrates for settlement or on a complex benthic community for certain types of food and shelter from predation. If habitat destruction imperils pre-harvest life stages, then the fishery is threatened by habitat destruction. Reserves are a primary means to protect such critical habitat, with the potential for enhancing biological productivity.

Box 2-1
Gag Grouper Case Study

The gag (*Mycteroperca microlepis*) grouper population in the Gulf of Mexico has 90% fewer males now than it did 30 years ago. What has happened to cause such a dramatic shift in the sex ratio?

Common fishing practices have disproportionately affected male groupers because of the complex biology and social systems that underlie this fish's reproductive behavior (Coleman et al., 1996). Groupers first mature as females; after receiving the appropriate social cues, some females become males. Aggregations are large groups of fish that form offshore for a relatively short but critical period during the spawning season. If dominant females encounter too few males in the spawning groups, they will change sex in the ensuing year so that more males will be available by the following spawning season. The spawning season and aggregation behavior are confined to a brief period in the late winter or early spring. At other times of year, males and females occur in separate locations, with males remaining offshore while females move to shallower water.

The grouper fishery targets spawning aggregations and the largest fish to obtain the highest yield for the least effort. Because males are larger and attack bait more aggressively, they are caught more frequently. At the same time, the population is less able to compensate for the disproportionate loss of males because fishing disturbs behavioral interactions that promote female-to-male sex change, the natural mechanism for maintaining a favorable sex ratio.

How widespread is the grouper overfishing problem? Declines in the abundance of males have been associated with exploitation in the Gulf of Mexico of both gag (*M. microlepis*) and scamp (*M. phenax*)—closely related species with similar life-history characteristics. Declines also have been reported for Atlantic stocks of gag and scamp, as well as other grouper species (e.g., Shapiro, 1979; Nagelkerken, 1981; Bannerot, 1984). In the southeastern United States, 11 of the 19 most important reef fish species are overfished or on the verge of being overfished. Most of them are groupers, and all groupers change sex from female to male. At least two suffer from low proportions of males in the population, and all of them aggregate to spawn. Currently, depleted Nassau grouper (*E. striatus*) and jewfish (*E. itajara*) are completely protected from fishing in state and federal waters. Two more species, Warsaw grouper (*E. nigritus*) and speckled hind (*E. drummondhayi*), can no longer be commercially fished or sold, and the recreational catch is limited to one fish per vessel per day. Finally, 26 grouper species worldwide are being considered for listing as vulnerable to extinction by the International Union for the Conservation of Nature and Natural Resources.

Current fishing regulations are insufficient to preserve either the social structure or the natural proportion of males among these fishes. Most management approaches fail to address the critical aspects of grouper reproduction. Establishing reserves at spawning aggregation sites could serve four functions:

1. Protect highly vulnerable aggregations from concentrated fishing effort, thus distributing effort over larger areas and longer periods of time.
2. Protect spawning fish so that spawning activity is not disrupted.

(continues)

Box 2-1
Continued

3. Prevent disruption of the normal behaviors and social interactions that trigger sex change.
4. Selectively reduce fishing mortality on males.

The concept of designating reserves to protect spawning sites is beginning to influence the management of grouper stocks. In June 2000, the National Marine Fisheries Service approved closure of two areas to fishing (except for highly migratory species) totaling 219 square nautical miles[a] in the northeastern Gulf of Mexico. These closures will be used for scientific evaluation of marine reserves, both to protect spawning aggregations of gag and other groupers and to evaluate the effectiveness of reserves in maintaining a more balanced sex ratio by protecting male gag from excessive fishing pressure.

---

[a] Throughout this report, nmi[2] will be used in place of square nautical miles.

Fisheries also impact other nontarget species that are taken as bycatch and often discarded dead at sea. Some fisheries have to be shut down before their quotas can be fished because bycatch limits for nontarget species are reached first. For example, regulations established by the North Pacific Fishery Management Council to reduce bycatch mortality of Pacific halibut (*Hippoglossus stenolepis*) in Alaska frequently lead to closing the groundfish fisheries for Pacific cod (*Gadus macrocephalus*), rock sole (*Pleuronectes bilineata*), and yellowfin sole (*P. aspera*) when the halibut mortality cap (currently set around 7,000 megatons [mt]) is reached (Adlerstein and Trumble, 1998). Groundfish fisheries suffer because they cannot catch their quotas, and the directed halibut fishery suffers because halibut recruitment is reduced and the biomass of adult halibut killed as bycatch is deducted from the allowable commercial catch. A total annual yield loss of about 11,000 mt was estimated at a time when the halibut directed catch was close to 32,000 mt (Clark and Hare, 1998). Reserves placed in areas where catch rates of nontarget species are persistently high may significantly reduce bycatch rates (i.e., mortality of nontarget species per unit of target species caught) and alleviate some of the problems of multispecies fisheries. Certainly, reserves are not the only means to control bycatch problems; in fact, depending on the situation, gear modifications and other management tools (e.g., individual bycatch caps, mandatory landing of all bycatch species) may prove more effective.

Ensure Against Possible Failures of Conventional Regulatory Systems

Because stock assessment methods can be inaccurate (NRC, 1998a), especially given the limitations of the data normally available for assessments (NRC,

2000c), reserves have been proposed as a way to ensure that harvest rates will not exceed some maximum bound or stock levels will not fall below some minimum threshold. Many managed fisheries in North America are regulated by placing annual limits on the total catch of individual species. Alternatively, the amount and quality of fishing effort are controlled to achieve the desired target harvest rates. However, these methods do not always work, particularly in the absence of reliable estimates of stock biomass (for catch quota systems) or of fishing mortality and its relation to the amount of fishing effort (for effort control systems).

Conserve Life-History Traits and Genetic Diversity

Most fishing methods are strongly size selective, commonly removing the largest and oldest fish at a higher rate (e.g., Parma and Deriso, 1990). This may exert strong directional selection toward slower growth and smaller size at maturity (Ricker, 1981; Thorpe et al., 1983; Policansky, 1993). Sex ratios can also be significantly skewed as a result of fishing when one of the sexes is differentially removed. Sequential hermaphrodites are a classic case; for example, many exploited stocks of groupers change sex from females to males, and the proportion of males in the stock has been critically reduced by fishing (Box 2-1; Coleman et al., 1996). More generally, differential mortality by sex may be due to large sexual size dimorphism or peculiarities in the mating system that result in one sex being more vulnerable than the other. By relaxing the selection pressure from fishing in some segments of the populations, reserves may help conserve the natural genetic diversity for life-history traits (Trexler and Travis, 2000).

Scientific Knowledge

Provide a Source of Baseline Data

Marine ecosystems are highly variable associations as a result of both natural variation and anthropogenic effects. Because all factors effecting change in ecosystems operate simultaneously and at different temporal and spatial scales, it is extremely difficult to discern natural from human-induced causes, and this is probably not possible without representative baseline studies and benchmark, undisturbed habitats to use as standards in the evaluation of human-induced impacts. Marine reserves offer the only means of protecting such baseline sites in areas that are otherwise affected by human activities.

Understanding fish population dynamics is hindered by the difficulty of separating fishing effects from natural environmental variability. In species that have low mobility, fishery reserves would provide an unfished control to compare population dynamics inside with dynamics in areas under conventional management. In many cases, stocks managed as separate units might be interconnected through larval and juvenile dispersal and thus would not be dynamically separate replicates. However, even in these situations, protecting some stock subunits from fishing may facilitate research on postdispersal processes (e.g., recruitment and growth) and how they are affected by local density and changes in habitat structure from fishing. Reserves could serve an important role in fisheries research as a tool to study fishing impacts through spatially replicated areas under different management regimes.

Educational Opportunities

MPAs provide a unique opportunity for the public to learn about the diversity of marine life and how human activities both on land and in the sea affect the health of marine environments. Many MPAs, like parks on land, establish interpretive centers and prepare educational material for schools and recreational groups.

Enhancement of Recreational Activities and Tourism

Tourism and recreation could contribute significantly to the commercial value of an MPA. The aesthetic appeal of marine areas for tourism is dependent on the quality of the natural environment—abundant marine life, unpolluted waters, intact habitats. An MPA may serve as a catalyst for the development of a suite of nonconsumptive services that include such diverse elements as shore-based aquaria and museums and seagoing activities such as whale watching. Recreational activities that do not threaten the protection of marine life not only provide local communities with economic opportunities, but also may enhance appreciation and support for the MPA.

Sustainable Environmental Benefits

Marine ecosystems provide benefits beyond harvestable products such as fish and algae. Sometimes referred to as a category of ecosystem services, these benefits include processes such as water purification, protection of coastal areas from storm damage (coral reefs, mangroves, seagrass beds), bioremediation of chemical and oil spills, reduction of atmospheric carbon dioxide through biological carbon sequestration, and nutrient cycling. MPAs and reserves can support the maintenance of marine ecosystems and the services they provide.

Protection of Cultural Heritage

MPAs can also be established to protect areas of distinct character with significant cultural value. Examples of these are protecting archaeological sites, shipwrecks, places of special historical significance, and landscapes or seascapes to assure the continuation of traditional uses, cultural practices, and sacred sites. These areas fall under Category V of the IUCN system (see Appendix F). The oldest national marine sanctuary in the United States, for example, was designated to protect the site where the Civil War vessel *Monitor* sank in 1862.

SUMMARY

Most MPAs will be established to fulfill several of the goals described above. The purpose of this report is to examine the potential of MPAs, especially areas zoned as marine reserves, for achieving the goals of preserving biodiversity and improving fishery management.

3

Conventional Management
of Marine Fisheries

PROBLEMS AND ISSUES IN FISHERY MANAGEMENT

Current interest in marine protected areas (MPAs) reflects dissatisfaction with conventional approaches to the conservation of marine ecosystems, especially fishery management, which often have failed to meet societal goals for sustainable use of marine resources and protection of biological diversity and productivity. Overfishing represents one of the most challenging problems in marine conservation. On a global basis, 44% of the world's fish stocks are now fully exploited, and 25% are overexploited and clearly in need of urgent conservation and management measures (Garcia and Newton, 1997; FAO, 1999). Collapses or dramatic declines of marine fish stocks, for example the Atlantic cod (*Gadus morhua*) off Newfoundland and on Georges Bank, call attention to the potential for failure resulting from the limitations of fishery science and of the current management system (Walters and Maguire, 1996; Fogarty and Murawski, 1998). The view of fishery scientists is that it was partly a failure of science that caused the collapse of the Newfoundland cod. Scientific advice to managers was not always correct or timely, and management failed to act in time on either the erroneous advice or the corrected advice. In the United States, the overexploitation of bluefin tuna and of mixed-species groundfish stocks convincingly illustrates the consequences of overcapitalization and excess effort.

Problems associated with conventional fishery management that threaten the sustainability of marine fisheries have been highlighted in recent reviews (e.g., NRC, 1999a). Overfishing and attendant fishing mortality rates that are too high and poorly regulated lead the list. Overcapacity, the presence of too many partici-

30

pants or units of effort in fisheries, is a related and serious problem. Open access to fisheries tends to favor overexploitation rather than stewardship, a common impediment to effective management (NRC, 1999b), although very few commercially important fisheries in North America and Europe are completely open access in the sense that most are regulated, some have limited entry, and many have restrictions on effort.[1] Another problem is the failure of management to act expeditiously and conservatively or to respond appropriately. Ecosystem-based approaches to management (e.g., NMFS, 1999) have emerged from concern about bycatch, habitat destruction, and the failure to consider important biological interactions (e.g., predator-prey). Shortfalls in the ability of scientists to produce accurate stock assessments have at times provided poor advice to managers (NRC, 1998a, b). Stock assessments and resultant management measures always contain a level of uncertainty. To be effective in the face of this uncertainty requires that the assessments be interpreted conservatively so that stock size is not overestimated and subsequently overfished. Conservative, flexible, and adaptive approaches can compensate for the uncertainty of stock assessments, but frequently these features are lacking from conventional management.

Overfishing is in large part a consequence of excessive effort and capacity in fisheries (NRC, 1999a). Too often, fishery managers have been unable to control fishing effort, resulting in unsustainable levels of catch. This has been a particular problem for open-access fisheries where management does not limit the number of participants or high individual effort (see Chapter 4). In this situation, the economic incentives favor short-term exploitation over long-term sustainable use because the economic benefits of sacrificing current catch to rebuild the stock are intangible compared to short-term needs (bills to be paid), and long-term benefits may have to be shared with newcomers when the fishery recovers (Hilborn et al., in press). As more people enter the fishery or improve their fishing capabilities, the future yield to the individual fisher decreases. This often fosters competition to maintain or even increase individual catch levels even as stocks decline. In response, managers may shorten fishing seasons; participants then increase their fishing power, and effort becomes concentrated in time, sometimes resulting in "races for fish" or "fishing derbies." In the worst cases, derbies are absurdly brief, lasting only two days in the Pacific halibut (*Hippoglossus stenolepis*) fishery in the United States during the early 1990s before individual quotas were implemented (NRC, 1999b).

In addition to depletion of fishery stocks, there are unintended consequences of fishing, such as bycatch and degradation of habitat from destructive fishing practices (Dayton et al., 1995; Watling and Norse, 1998). Bycatch here refers to

---

[1] Open access is defined as the condition in which access to a fishery is in effect unrestricted (i.e., no license limitation, quotas, or other measures that would limit the amount of fish an individual fisher can land) (NRC, 1999b).

the incidental catch by fishing gear of adult and juvenile fish that are not the target of the fishery (Alverson et al., 1994). Bycatch and habitat loss not only may have deleterious effects on fishery yields, but also may degrade the ability of marine ecosystems to support biological diversity. Therefore, effective regulation of fishing activity in the oceans is not just a fishery management issue. For example, unique features and habitat such as coral reefs need prohibitions on fishing, as well as protection from shipping, diving, recreational boating, and destructive coastal development. Ecosystem approaches, including marine reserves, will have to be added to the conventional management toolbox to conserve biodiversity, maintain biocomplexity, and ensure that ecosystem services are maintained for posterity. The public's interest in ecosystem approaches in part represents the existence values that the public places on preserving the diverse biota and habitats of the sea (see Chapter 4). To ensure the future of living resources and habitats in many stressed marine ecosystems, some areas of the ocean could be zoned in MPAs for limited access and use. This chapter describes conventional fishery management tools, noting both limitations and failures, to provide a context for evaluating MPAs and reserves as complementary or alternative tools.

CONVENTIONAL FISHERY MANAGEMENT

In general, conventional fishery management seeks to maintain high, yet sustainable, yields by regulating the number or weight of fish caught, the size of fish caught, or the time and space (area) within which fishing is allowed. The intent in each case is to control fishing mortality rates. Conventional approaches to fishery management in the United States can be succinctly characterized by three main components: (1) an underlying fishery science and management paradigm, (2) a set of conventional management tools, and (3) the fishery management system.

Fishery Paradigm

Fishery management relies on estimates of the population size of a target species to determine how many fish or what fraction of the population's biomass can be caught without damaging its reproductive potential. To make these determinations, management depends on a conceptual model of a fishery that makes three simplifying assumptions: (1) the fishing fleet targets and exploits a single-species stock, (2) the stock of interest is segregated temporally or spatially from other stocks, and (3) the individuals are perfectly mixed so that the effects of fishing are well spread over the whole stock. These assumptions, which are far from true in most situations, can have serious consequences for the effectiveness of fishery management.

Single-Species, One Stock

Most management measures are directed at individual stocks of a single species and do not take into account species interactions, such as predator-prey relationships. A basic assumption of most models used to determine a catch level is that the catch rate a stock can sustain can be designated based upon the average productivity of the stock. Productivity, in turn, is presumed to depend primarily on the size of the adult stock. In this scenario, controlling adult stock size is the primary means of ensuring sustainability of the fishery. Furthermore, stocks are assumed to respond in a density-dependent manner and therefore are postulated to have maximum productivity at intermediate stock sizes. Thus, maintaining the stock size that allows maximum sustainable yield (MSY; see Appendix B for definition) historically has been a major management goal, and fishing at a rate that produces MSY on average over good and bad years has been the target. In fact, fishing at the MSY level (a fixed exploitation rate policy) does not ensure constant catches in the future or a stable adult population size because of substantial variability in reproductive success and recruitment. It was recognized more than two decades ago (see Larkin's, 1977, famous epitaph on the concept of MSY) that it is therefore too risky to set a constant quota for catch at MSY. In good years, fishers may prosper with MSY-based catches, but in years when the environment is less favorable and recruitment and productivity decline, the stock will diminish and MSY may quickly lead to overfishing. The fishing rate corresponding to MSY (F_{MSY}) still remains a criterion in determining optimal yield, the regulatory target used to manage marine fish stocks in the United States (NOAA, 1996a). However, F_{MSY} now is viewed by many as a threshold that should not be exceeded, rather than as a target at which to aim. More conservative quotas and exploitation rates are now recommended, due in part to recognition of our limited ability to estimate and implement F_{MSY} or, for that matter, other target fishing mortalities via catch or effort control. As Hilborn and Walters (1992) noted, obtaining an estimate of MSY (or F_{MSY}) usually requires fishing at levels that already exceed it.

Fish Stocks That Are Temporally or Spatially Segregated

Although it is obvious that management must be tailored to individual species' life histories, the individual stock and single-species approaches to management are ineffective for multispecies or even mixed-stock associations, in which many different species or stocks with similar habitat and prey requirements overlap in their ranges. Good examples are reef fish off the southeastern United States and rockfish in the northern Pacific Ocean. Warm-temperate species such as gag (*Mycteroperca microlepis*), scamp (*Mycteroperca phenax*), and red hind grouper (*Epinephelus guttatus*), for instance, co-occur to such an extent that catch restrictions placed on one species in the complex typically result in increased regulatory discards while fishing for associated species. The same is true for the 83 species of rockfish managed as a complex off the Pacific west coast. Inevitably, regulatory discards will increase the mortality of the restricted species and threaten its recovery.

Individuals Are Perfectly Mixed

A key element of the fishery management paradigm is the concept of a well-mixed stock. Migration patterns and more general spatial processes are fundamental components of fishery science. Knowledge of spatial processes serves to delineate management units, each viewed as a "dynamic pool" isolated from the rest. Most theory and management have been conceived for large-scale, commercial fisheries that target relatively mobile species (e.g., tuna, plaice, gadoids), for which dynamic pool assumptions may provide a reasonable simplification, at least at the scale of a fishing ground. In this conceptual model, because the effects of fishing are "diluted" in the pool, the use of spatially explicit approaches to manage each unit has been largely missing (Shea et al., 1998). This conceptual model has dominated marine fishery management. As a result, it has been applied indiscriminately to low-mobility species for which the paradigm is clearly inappropriate, such as rockfish (*Sebastes* spp.) and sedentary invertebrates. Fishing effort on relatively sedentary species preferentially targets the highest fish concentration and results in a mosaic of fishing mortalities and serial depletion of fishing grounds. Management by conventional means is complicated because the dynamics of the resource, fishing process, and monitoring are dominated by local processes that are often impossible or impractical to assess. In reef fisheries, for example, the relevant scale for assessment and management may correspond to a single reef.

Another tool to limit fishing effort is control of access to the fishing grounds. This strategy, known as "spatial management," can be applied as either temporary (seasonal or year-to-year) or permanent closure of portions of the fishing grounds. Spatial management can be used to control access to resources and probably has been practiced for centuries (Cushing, 1988), but it has not played a central role in the management of marine fisheries. Its importance in the management of benthic shellfish is now accepted (Orensanz and Jamieson, 1998; Perry et al., 1999), as are novel management schemes involving area rotation (Bradbury 1990, 1991; Perry et al., 1999). However, many more species could benefit from spatially explicit management, particularly those with a relatively stationary adult stage. A major constraint on implementation of spatial management is the lack of spatial catch data for many species and the complexity of spatially explicit stock assessment models.

Although spatially explicit components usually are missing, some existing fishery models *do* recognize that variables other than adult stock size affect productivity. However, these variables usually are treated as random variations beyond human control. Consequently, management focuses on regulating the size of the catch and the effort directed at obtaining it, while environmental factors are downplayed. Even when not controllable, persistent trends or variations in productivity driven by environmental conditions should be considered in policy evaluation. In addition, environmental variables that affect habitat quality and are directly affected by human activities are now addressed explicitly in fishery management plans, following the reauthorization of the Magnuson-Stevens Fishery Conservation and Management Act (MSFCMA, NOAA, 1996a) and its emphasis on essential fish habitat (EFH). This focus on habitat has initiated a shift toward spatial considerations and designation of areas important for the productivity of economically important species. These designations, of essential fish habitat and habitat areas of particular concern, have bolstered interest in protected areas and supported the use of MPAs as a legitimate tool for fishery management.

Long-term, large-scale closures have been instituted as single-species refuges from fishing to promote rebuilding of depleted stocks. In many respects, they resemble permanent reserves, except that such areas may revert to their former status when restoration is attained. Long-term closures are becoming more common as a population-rebuilding tool (e.g., northern cod stock off Newfoundland), in which depleted populations may require many years to restore. Long-term closures might achieve some goals of reserves, although benefits may be transient if subsequent fishing mortality cannot be controlled.

Another form of spatial management is the use of rotating fishing areas. Here only a fraction of the fishing grounds is opened in a given season, the rest being closed for specified periods (often years) to promote growth of young animals and allow them to reach more valuable sizes. The area opened is rotated from year to year. This approach combines temporal and spatial closure to regulate fishing. In addition to rebuilding fish stocks, rotating closures may allow habitat and biological communities to recover from the effects of fishing, such as damage to bottom habitats by trawling and movement of "fixed" gear. However, recovery of habitat and biological communities may require closures on the order of 5 to 10 years (Collie, 1997).

Conventional Management Tools

Management based on the fishery paradigm above centers on measures that regulate fishing activities and the level of catch, rather than on measures that directly promote management of habitat or consideration of environmental variables affecting fish productivity. Generally, the goal is to manage exploited populations such that they are maintained at productive levels (close to or above MSY) that support a high yet sustained fishing yield and to require rebuilding plans when yields fall below a minimum stock size threshold.

Conventional management approaches to control exploitation rates fundamentally rely on placing limits on the amount or efficiency of fishing effort (effort-controlled fisheries) or on the total amount caught by specifying catch quotas and allocations (quota-based fisheries). In both cases, a target fishing rate is first specified, whether constant or variable in response to stock condition, based on analysis of historical experience with the fishery, on experience with fisheries for similar species, or on modeled responses of the fishery to simulated fishing mortality.

Effort Controls

There are many forms of effort controls, including restrictions on gear, vessels, time fished, and number of fishers. These are usually the first controls applied to a fishery to slow the rate of catch. Gear restrictions can include the type, amount, or dimensions of gear or specific features of the gear such as net mesh size, hook spacing on longlines, or configuration of fish traps. Vessel restrictions may include design, length, or engine horsepower. The number of fishers can be regulated by allocating licenses to either fishers or vessels. Time fished can be regulated through limiting the amount of time available for fishing through seasonal closures, "days-at-sea" restrictions, or specific days and hours when fishing is permitted.

To implement a target fishing rate by means of effort regulations, managers need a reliable estimate of the fishing mortality caused by each unit of fishing effort to be allowed, a parameter known as *catchability*. The estimate of catchability is then used to determine the amount of fishing effort (e.g., the number of total days at sea to be allowed) that is compatible with a chosen exploitation rate or fishing mortality so that

$$F_{target} = \text{catchability} \times \text{effort}.$$

Catchability is generally assumed to remain constant as stock size varies. However, catchability may change—for instance, fishers may increase their efficiency when the stock declines—and this could result in overexploitation, a common reason for the failure of conventional management.

In addition to measures that attempt to regulate exploitation rate directly, other forms of effort control may be implemented to reduce fishing power, protect vulnerable life-history stages, or increase the market value of the fish. Temporal closures, for example, are frequently used to protect fishery resources at times when they are particularly vulnerable to fishing, such as when fish aggregate on spawning grounds. Also, temporal closures can be used to allocate fishing over the season in a manner that increases the value of landings. For example, fishing might be prohibited during parts of the year when the population is composed of small or poorly conditioned fish. Closure of the Dungeness crab (*Cancer magister*) fishery during the molting season when meat quality is poor provides one example (Methot, 1986). Broader application of temporal closures to protect whole communities or complexes of species and habitats is less common, but perhaps of greater relevance to developing marine reserves as an ecosystem approach to fishery management.

Catch Controls

A common approach to controlling fishing is to regulate the catch or the amount of fish landed. This is the favored method used to regulate fisheries in Alaska and along the west coast of the United States. Quota-based management relies on the ability to model relative trends in abundance over time and to estimate the absolute size of the exploitable stock. This information is used to set the total allowable catch (TAC) that meets a chosen target exploitation rate.

TACs or quotas are typically calculated as the product of a target exploitation rate μ and an estimate of current stock biomass $\hat{B}_t$:

$$TAC_t = \mu_t \, \hat{B}_t.$$

In principle, TAC-based management can be a direct, efficient way to limit catches. However, the success of TAC-based systems depends on accurate estimates of stock abundance and biomass. Because the required level of accuracy is usually not available, the risk of overfishing may be high (Walters and Pearse, 1996; Walters, 1998).

In many heavily exploited fisheries, both effort and catch controls are used to manage the fishery. In the Pacific halibut fishery, for example, a catch quota is used to control the exploitation rate, there is a minimum size limit on landed fish, and all fishing methods except setline gear are prohibited for the directed fishery (http://www.iphc.edu). Seasonal closures are in place, which prevent the interception of fish from different regulatory areas when the fish migrate from feeding to spawning.

Both forms of regulation, catch and effort controls, have significant shortcomings because both depend on the quality of stock assessments. In turn, the accuracy and reliability of stock assessments depends on data that are frequently limited or unavailable. Conventional methods used to estimate stock abundance and current rates of fishing mortality require good historical catch statistics under significant levels of exploitation and indices of stock abundance that reliably show population trends. Numerous methods are used for stock assessment (reviewed in NRC, 1998a); most are based on analyzing the way abundance changes in response to known catch levels. In addition to estimates of stock abundance or fishing mortality, conventional management depends on knowledge of what exploitation rates are adequate to derive "biological reference points" that designate both threshold and target levels of biomass and fishing rates. The sophistication of procedures used can lead to overconfidence in their ability to estimate abundances of stocks and their resilience to fishing pressure. This misplaced confidence contributed to the collapse of the Newfoundland cod fishery (Walters and Maguire, 1996). Although stock assessments usually are done competently by fishery scientists in the United States, the statistical uncertainty associated with estimates and biological reference points can lead to failed management (NRC, 1998a).

Fishery Management Systems

Management of fisheries in the United States typically is undertaken at geographic scales that range from local to national. Assignment of responsibilities and implementation of effective management is complex. Jurisdictions of responsible institutions and agencies may overlap in some fisheries. The eight regional fishery management councils (NOAA, 1996a) have primary responsibility for management in the U.S. exclusive economic zone (EEZ), but they may share responsibility with other regional management institutions for coastal migratory species, especially those that occur in the nearshore and estuarine regions of the coast. For example, the Mid-Atlantic Regional Fishery Management Council shares responsibility for managing coastal species such as bluefish (*Pomatomus saltatrix*), weakfish (*Cynoscion regalis*), and summer flounder (*Paralichthys dentatus*) with the Atlantic States Marine Fisheries Commission, which represents state interests in migratory species that are fished in the coastal zones and estuaries of Atlantic Coast states. Management systems are even more complex for such species, because state agencies also are engaged in the regulatory process within their jurisdictions. Furthermore, management of these species is conducted at additional regional levels (e.g., the Chesapeake Bay, in which the States of Virginia, Maryland, and Pennsylvania; the Potomac River Fisheries Commission; and the District of Columbia exercise jurisdictional control).

Management systems typically institute a variety of output and input controls to regulate fisheries over their geographic ranges. Quota allocations, often among sectors of the fishery (e.g., commercial and recreational), are common; minimum sizes or other size regulations may apply. Restrictions on gears, seasons, seasonally closed areas, and combinations of methods, often with specific geographic regulations within the range of the targeted species, are the tools that managers commonly apply. Not only is it difficult to attain consensus to manage resources, but the success of management measures is often uncertain.

The uncertainties in the success of management systems lie in the attendant uncertainties that usually characterize the science and management of fishery resources. The science of stock assessment itself may be uncertain for many fished stocks. Political pressures on managers and institutions can dictate management policies and responses, sometimes to the disadvantage of long-term benefit to fisheries. Disputes among sectors of fisheries—for example, different gear users, or recreational versus commercial fishers—can dominate the dialogue and sometimes result in compromises that do not constitute best management policy. Faced with uncertainty in science and social conflict, managers historically have been slow to act to conserve fishery resources. Legislation, such as the national standards of the MSFCMA (NOAA, 1996a), presents, at least to some, conflicting goals of conservation, economics, and social interests that delay or misdirect management actions. Finally, to be effective, management systems must encourage compliance, either through enforcement or by providing proper institutional incentives to comply with regulations. As discussed in Chapter 4, management systems that confer user rights and participation of stakeholders in the management process can improve compliance.

As noted above, management in most of the U.S. EEZ is regulated by regional management councils (NOAA, 1996a). Currently, there are 37 fishery management plans submitted by the regional councils and approved by the Secretary of Commerce. The eight councils have jurisdiction over broad, discrete geographic areas, although sometimes their management authority is shared for migratory species. Jurisdictional issues may be significant for migratory stocks, especially coastal stocks that cross the boundary between state and federal waters at 3 miles from shore (for most states). It is important to note that if fishery reserves become an important and integral part of management plans, state and coastal regional authorities (e.g., the Atlantic States Marine Fisheries Commission and its Gulf of Mexico and Pacific Coast counterparts) will have shared, possibly complex, jurisdictional authority for spatial management and enforcement. At the time of this report, some of the regional management councils are considering and developing strategies for fishery reserves and other spatially restricted fishery management plans.

In some fisheries, managers have adopted methods that control access by establishing individual fishing quotas (IFQs) (NRC, 1999b), which assign shares of the fishery-wide TAC to selected individuals or sectors of a fishery. The privileged access that is afforded by IFQ management has been criticized by some, but it represents a step by conventional managers to match capacity and effort with available fish. Assigning rights or privileges to access is not, of course, sufficient to manage fisheries unless additional conventional tools of fishery management are also applied, such as quotas and gear restrictions, and special attention is given to controlling bycatch and discards, which can be problematic in IFQ fisheries.

New paradigms are emerging to guide management of marine fisheries in the new millennium. Although many of these paradigms build on conventional management practice, they have significantly changed the philosophy of management agencies in the past decade. The precautionary approach and the risk-averse policies that it implies have been advocated globally (FAO, 1995) and in the United States (NOAA, 1996a, 1999; NRC, 1999a; Restrepo and Powers, 1999). The burden of proof is being shifted away from demonstrating a negative effect of fishing before curtailing effort, to demonstrating that fishing practices will not damage the stock, habitat, or other ecosystem properties before allowing fishing to increase (Dayton et al., 1998). Although progress is slow, management is moving toward multispecies approaches, and ecosystem approaches eventually may be widely applied in managing marine fisheries (NMFS, 1999). Finally, the concept of embedding fishery management in the broader context of coastal zone management is being debated. It is here that MPAs can make an important contribution to accomplishing integrated management of our nation's coastal resources.

Long-term, single-species area closures represent a move toward MPA-style management. Although they have some features in common with reserves, single-species closures lack many key conservation benefits of permanent reserves and their objectives are generally narrowly drawn. Few temporal closures are designed to address multispecies or ecosystem concerns; rather temporal closures are a tool for single-species fishery management. An exception is the closure of areas 10-20 miles offshore of haulouts and rookeries occupied by endangered Steller sea lions (*Eumetopias jubatus*) to fishing for walleye pollock (*Theragra chalcogramma*). Other time and area restrictions have been implemented for the pollock fishery within and outside critical habitat for Steller sea lions.

UNCERTAINTY, FISHERY MANAGEMENT, AND A ROLE FOR MARINE RESERVES

Many scientists believe that a primary cause of fishery management failures is the inherent uncertainty in stock assessments. This uncertainty contributes to ineffective or untimely management actions and the reluctance of fishers to accept the economic costs of reducing effort even when stocks are in decline or their status is uncertain (Ludwig et al., 1993). To provide insurance against stock collapse, scientists have proposed establishing fishery reserves when the lack of accuracy in stock assessments and lack of resolve to fish conservatively make it difficult to achieve sustainable fishing levels under conventional management. The specific causes leading to the collapse of a fishery are controversial because it is difficult to discern the relative contributions of fishing pressure and environmental forces. Also, management generally does not account for the effect of environmental degradation on MSY (e.g., Myers et al., 1996, 1997; Orensanz and Jamieson, 1998; Caddy, 2000). Fishing fleets are ever more efficient at locating and catching remaining fish aggregations, with the result that once the fishery collapses, it may require long periods of time to recover, on the order of a decade or more, even in the absence of fishing (Hutchings, 2000). Ensuring against collapse is a primary but elusive goal of marine fishery management.

Central to the problem of uncertainty in fishery science and management is our difficulty in confronting it. Conventional fishery management relies on science, particularly our ability to determine appropriate target catches and to estimate actual fishing mortality or stock size as a basis for recommending effort or catch controls to meet these targets. Even when science is adequate, the effectiveness of management in achieving the desired control (i.e., control the exploitation rate) may be uncertain (Walters and Parma, 1996; Walters and Pearse, 1996; Walters, 1998). Experience and simulation analyses have shown that stock assessment methods sometimes are prone to errors exceeding 50%, even when costly monitoring programs are in place (NRC, 1998a). Worse, errors tend to be correlated from year to year, compounding their effects over time. Retrospective analysis often reveal biases, with stock size initially overestimated or underestimated for several consecutive years (Sinclaire et al., 1991; Parma, 1993). When scientists and managers depend on catch data from the fishery itself (i.e., fishery-dependent data), levels of bycatch and discards at sea often are unknown, and these sources of fishing mortality may not be included properly in assessments. Fundamental parameters, such as the rate of natural mortality, can be specified only in a rather broad range, based on life-history correlates. Indices of abundance derived from research surveys are valuable, but they too can be imprecise or, in many fisheries, simply unavailable.

It has been argued (Walters and Pearse, 1996; Lauck et al., 1998; Walters, 1998) that uncertainty in stock assessments is simply too large to manage fisheries sustainably using conventional tools. Three main approaches have been proposed to address this uncertainty: (1) choose substantially lower catch rates as fishing targets than in the past (Mace, 1994; Restrepo and Powers, 1999); (2) implement management tools that are less dependent on stock assessments, such as reserves (Roberts, 1997a; Lauck et al., 1998; Walters, 1998; Murray et al., 1999) and size limits (Myers and Mertz, 1998), and (3) generate institutional incentives that encourage responsible behavior on the part the fishers, such as different forms of user rights (NRC, 1999b; Hilborn et al., in press). These three approaches are not exclusive, and all may have to be considered for fishery management to be successful. Marine reserves, as an alternative to conventional management, also have uncertainties associated with their performance. Sources of costs and benefits of some of these approaches are discussed in more detail in Chapter 4.

4

Societal Values of Marine Reserves and Protected Areas

Designating a significant amount of coastal regions as marine protected areas (MPAs) and reserves is likely to alter both the kinds of benefits or ecosystem services provided by the marine environment and the distribution of these benefits among different groups and individuals. Because the United States government has *public trust* responsibilities to manage federal waters for the interests of citizens nationwide, assessment of the various costs and benefits of establishing MPAs requires evaluation of public opinion from both direct users and citizens concerned about marine conservation. The acceptability of MPAs to the general public and to direct users will depend significantly on whether the perceived benefits are greater with or without MPAs, and this, in turn, will influence the political support for MPA programs.

All marine systems provide a range of benefits to humans, even if their resources are not exploited. These benefits span a spectrum from direct on-site user benefits to indirect benefits accruing to individuals who do not use the marine ecosystem directly. On-site user benefits are generally associated with consumptive uses (recreational and commercial fisheries; seaweed harvesting; shell, coral, and sponge collecting), but important nonconsumptive uses (tourism, diving, bird and whale watching, the aesthetics of natural areas) are also provided by marine ecosystems. Many of these on-site activities generate income directly to participants and indirectly to coastal economies that service the activities. Even more difficult to evaluate, but equally real, are the *heritage* or *existence values* associated with the public's appreciation of unique and natural systems. In addition, marine ecosystems provide hard-to-quantify off-site benefits as components of regional and global climatological, biological, and chemical systems, including removal of carbon dioxide from the atmosphere, production of oxygen, moderation of coastal temperatures, and powering terrestrial hydrologic cycles (Daily et al., 1997). This chapter describes these different types of values, the potential costs and benefits of MPAs in supporting these values, methods for evaluating societal values, and finally the need for community involvement in the decisionmaking process.

ORIGIN OF THE VALUES ASSOCIATED WITH MARINE ECOSYSTEMS

The "natural" functioning of marine ecosystems has included human influences for significant periods of time (Zacharias et al., 1998). In North America, coastal areas have been affected by human activities starting with the migration of people across the Bering Sea land bridge and colonization of the West Coast more than 10,000 years ago. When Europeans arrived in the Americas, they encountered marine ecosystems already shaped by human influence. Human exploitation of marine resources changes the structure of ecosystems through impacts on the food web and habitat. Yet access to and use of the sea also affect the structure of human societies and the evolution of their perceptions of the values provided by marine systems.

Because humans are so efficient in capturing fish and other marine species, the human role in the ecosystem may be considered analogous to that of a keystone predator (Castilla, 1993). The impacts on the structure of coastal marine communities can be direct, indirect, or subtle and are revealed when humans are excluded from the ecosystem, for example, after establishing an ecological reserve. However, human impacts are mediated by influences other than typical predator-prey interactions that reflect unique human social characteristics such as cultural traditions, economic conditions, and technological advances. Cultural traditions can be characterized in terms of environmental ethics and cultural landscapes as described below.

Environmental Ethics

Biocentric values—valuing nature for its own sake—are important for many people as a function of their beliefs about the proper relationships between humans and nature. These beliefs are critical for explaining the adaptations of human cultures to their local, regional, and world environments. A key question in characterizing environmental ethics is whether or not humans are perceived as a part of nature or separate from nature (McDonnell and Pickett, 1993).

Increasingly, people in many nations value the quality of the environment and recognize that animals and plants have the right to some measure of protection from human disturbance (Inglehart, 1990, 1991, 1997; Abramson and Inglehart, 1995). When polled regarding the "environment-versus-economy" balance, more than 50% of people chose environmental protection over economic benefits in each of 24 nations, except Nigeria, India, and Turkey (Dunlap et al., 1993). These international trends, reflecting preferences for improving environmental protection, suggest that public values worldwide may support ocean conservation measures such as MPAs, based on environmental ethics alone.

How do these attitudes apply to the specific case of conservation in marine ecosystems? Human populations with extensive experience in the use of marine resources often develop a conservation ethic regarding those resources. This ethic directly reflects three factors: (1) the perception of local populations that have special access rights and responsibility for local areas, (2) the environmental knowledge and lessons they have learned from past experience using these resources, and (3) the expectation that future generations will derive subsistence from the ocean just as past generations did. Conservation ethics have developed in coastal populations in as few as three generations (Stoffle et al., 1994b).

Public values can be influenced by organized and collective efforts of relatively small numbers of people. Groups with either an economic interest or a conservationist agenda exert political influence and play a role in developing public awareness and values concerning ocean resources. For example, SeaWeb (http://www.seaweb.org) sponsored a survey conducted by the Mellman Group that showed much support for ocean protected areas (76% in favor) but little awareness of the existence of the National Marine Sanctuary Program (34%). This mobilized ocean conservation organizations to undertake campaigns to increase public understanding of MPAs and the status of national marine sanctuaries. Similarly, groups with an economic interest, such as coastal developers and the fishing industry, seek to influence policy through public information campaigns and political lobbying. Often, public values do not get translated into action because these communities do not have the institutional capability to influence regulatory policies (McCay and Acheson, 1987; Ostrom, 1990; Gibson and Koontz, 1998).

There are also many examples in which societies have severely overexploited marine ecosystems, reflecting a variety of circumstances. Hence, even when a coastal community develops a conservation ethic, short-term exigencies, such as a severe economic depression or a radical shift in climate, can disrupt sustainable practices to provide for immediate needs.

People also consciously damage the natural resources they exploit. For example, if there are no special access rights or responsibility (a factor in the development of a conservation ethic as described above), individual economic incentives favor maximizing current yield, even at the expense of the long-term health of the resource, because the individual has no guarantee that others will not overexploit the resource and thus jeopardize future yields. This consequence of open access has been termed "the tragedy of the commons" (Hardin, 1968).

Damaging behaviors also may occur with new user groups who are not familiar with the marine ecosystem or who may have displaced previous local inhabitants whose knowledge is either unsought or unavailable (Agardy, 1997). In other cases, users of a marine ecosystem may not be the decisionmakers. For example, they may be employees of large companies that exploit marine resources and, as such, lack the authority to practice sustainable resource use.

Environmental ethics can be examined systematically as part of the assessment and evaluation of areas being considered for MPAs by studies of their distribution among various groups of stakeholders. Assessing the acceptability of an MPA requires studies of stakeholders, including *social collectives* and *groups,* as they exist at the local, regional, national, and international levels. *Social collectives* are assemblages of people who do not interact directly, but have similar social characteristics such as age, sex, or income and share a distinctive and common body of interests, values, and norms (Merton, 1957). In marine and coastal environments, social collectives might involve all of the tourists who regularly visit a marine park or individuals who access a sanctuary Web site to monitor its condition. *Social groups* are assemblages of people who interact socially, are clearly bounded, have symbols of membership, and tend to share a distinctive and common body of interests, values, and norms (Merton, 1957). In marine and coastal environments, social groups might include local fishers' organizations, dive clubs, and incorporated communities.

It is essential to acknowledge that various social collectives and groups may hold different, or unexpected, positions regarding marine protection due to their own unique set of environmental values and the way they prioritize these values. Kempton et al. (1995) found that environmental values in the United States are organized into coherent cultural models among different groups and that these values are useful for predicting responses to environmental issues. Significantly, environmental values have become integrated with core American values such as parental responsibility, obligations to descendants, and traditional religious teachings.

Cultural Landscapes

Human values associated with marine ecosystems are related to understanding the relationships between components of the ecosystem and processes of change that occur (Kempton et al., 1995). The idea of a "cultural landscape" provides a cognitive framework for understanding links between physical places and human values (Stoffle et al., 1997; Zedeno et al., 1997). The theory of cultural landscapes includes (1) places (called landmarks), (2) spaces between places, and (3) a relational pattern that integrates space and place. Places may contain culturally significant artifacts (such as shipwrecks), or they may be natural places that are culturally significant, like the Skoskomish Indians' origin place at the mouth of their river in Puget Sound. The literature on the meaning of place and space is well established (Tuan, 1996).

The federal government recognizes cultural landscapes as protectable by law and regulation. Cultural landscapes may receive special land management status and protection by being incorporated into the National Registry of Historic Places. Places within landscapes can also be nominated to the National Register and, during this process, are called traditional cultural properties. Nominations based on geographic location tend to focus on cultural areas such as historic trails, but space-based areas, such as the trail-like routes of the Underground Railroad, may also be nominated.

Marine cultural landscapes reflect the way humans use and value various ecological zones in the sea and along the coast. Some of these cultural landscapes will more or less reflect the geographic boundaries of marine ecosystems and the diversity within them. Some marine areas meriting protection may be landmarks within landscapes, manifested either as special topographic areas such as seamounts, coral reefs, and entrances to underwater canyons or as special hydrological places such as estuaries and upwelling areas that are especially productive. Evaluation of cultural landscapes will help inform the process of choosing MPA locations.

Studies of the cultural landscape of a proposed MPA site should include stakeholder and user groups associated with the site. Methods for gathering social and cultural information should include various instruments for assessing culture, cognition, and values, including detailed ethnographic surveys. These methods allow measurement of environmental values, cultural models of nature, the cultural significance of places, and the integration of places and intervening spaces into cultural landscapes. Geographic information systems (GISs) can be used to produce ecosystem-wide maps as data recording and analysis tools. Such systems could be used to integrate the results of interviews, cultural landscapes, and environmental characteristics (species distributions, topography, ocean features).

A scientific understanding of the social groups and collectives potentially affected by a proposed MPA is important in terms of identifying stakeholders, designing the potential MPA, and meeting legal and regulatory mandates. Social impact assessments are required under the National Environmental Policy Act and the Magnuson-Stevens Fishery Conservation and Management Act. When local communities constitute unique ethnic or racial entities, social assessment may be required under the National Historic Preservation Act.

COSTS AND BENEFITS TO USER GROUPS

The beneficiaries of MPAs may include individuals who value the naturalness of marine areas, tourists who want to see intact marine environments and the animals that live there, divers who seek thriving natural habitats such as coral reefs, and fishers who want higher long-term yields from more sustainable stocks of fish. Some of these values can be characterized to a greater or lesser extent in economic terms, for instance, how much a diver is willing to pay to see living reefs in a marine reserve versus degraded reefs in an unprotected area. Similarly, a fisher can calculate how much income he or she may lose when effort is displaced by a fishery reserve and weigh that against reduced variability of the catch and potentially higher yield if the reserve protects against overfishing. Other non-use benefits, such as heritage or existence values, are difficult to measure in economic terms, but are no less important for weighing the costs and benefits of marine reserves. Potential sources of costs and benefits of marine reserves, including market and nonmarket values, have been summarized by Hoagland et al. (1995) (Table 4-1).

Policy Context

Direct users of marine resources attain access to services provided by marine ecosystems through public policies that place conditions on access rights and set regulations for various uses. For example, nearly every coastal nation has instituted fishery regulations that, in principle, protect and sustain the economic benefits available to commercial fishers. Other public policies and regulations mitigate conflicts among different user groups (e.g., allocating particular fishing areas to particular gear types that would otherwise clash, prohibiting certain extractive and polluting activities that would reduce recreational uses). Both broad conservation and specific conflict mitigation policies determine the spectrum of ecosystem services available, as well as which user groups will have access to these services. The point is that regulations and public policy already determine, to a significant degree, the portfolio and distribution of services that are provided by marine ecosystems in their current state of use. Any policy changes involving MPAs will alter the mixture of services, the set of beneficiaries of those services, and potentially the level of benefits from these services.

Existing policies reflect past and present political interplay among various user groups, each vying for a stake in the use of marine systems that cannot satisfy every user's wants. Therefore, existing systems of regulations reflect the history of the tug-of-war among different groups and do not necessarily represent a coordinated management plan developed through rational processes. Generally, the more that the economic benefits from marine ecosystems are directly appropriable by individuals, the more likely are such individuals to develop organized and successful political interest groups that will lobby for legislation and rules benefiting the group.

In contrast, some beneficiaries of ecosystem services are typically underrepresented in the political system. These individuals are often those who benefit from *public good* services. Public good services accrue to everyone once provided, but they are not individually appropriable, so consumption by one person does not detract from the consumption of others. Examples might be basic scientific knowledge, the heritage value of unique ecosystems, or the beauty of undamaged seascapes. These types of services are generally underprovided in a

TABLE 4-1   Sources of Costs and Benefits of Marine Protected Areas

Benefit	Cost
	Purchase of land and facilities
Strengthens property or liability rights to a clean marine environment	
New or improved opportunities:	Forgone opportunities:
Tourism, diving, boating	Mineral ED&P
Recreational fishing	Waste disposal
	Commercial fisheries
	Treasure salvage, shipping, tourism
Facilitates natural resource management	Administration
Rare ecosystems, species, stocks, cohorts, habitat, refugium	Monitoring and enforcement
Facilitates cultural resource management	Administration
Archaeological study, resource protection, recreation "targets"	Monitoring and enforcement
Oceanographic research	Research and education costs
Control area, ecosystem studies, public education	
Positive external effects	"Paper park":
Buffer zone, increased assimilative capacity, onshore development opportunities	Benefits small or nonexistent and industrial development opportunities forgone
Prevents development that is costly to reverse	Results in zoning decision that is costly to reverse
Nonmarket benefits	Nonmarket costs
Option—vicarious	Option
Bequest—existence	
Conceptual simplicity of boundary	Economic aspects of size rarely considered

SOURCE: Hoagland et al., 1995.

mixed public-private system because it is difficult to mobilize the constituency whose interests are at stake to a level that actually reflects the strengths of those interests (Samuelson, 1954; Olson, 1965; Starett, 1988). In part, this is because the benefit to any single person may be relatively small, although the cumulative benefit is large. Environmental advocacy groups often lobby as representatives of individuals who would benefit from the provision of public good environmental services. These groups have recently become more vocal and successful in the political process, manifesting a shift in public environmental priorities.

Current user groups frequently claim rights and protections to the use of living marine resources, analogous to the homesteading farmer's title to land, because use creates a source of income and wealth. On land, customary use has in some circumstances been converted into titled property rights. However, marine resources within federal waters, except usual and customary use associated with tribal rights, are held in public trust for the citizenry of the United States; for instance, no individual property rights exist for fish stocks, although individual transferable quota (ITQ) rights have taken on many of the attributes of individual property rights (NRC, 1999b). Furthermore, the government has "the right and duty to protect and preserve the public's interest in national wildlife resources."[1] Still, the adoption of marine reserves on a large scale would be viewed by current users as a change in the right of access to natural resources with consequences for the value of investments made in vessels and gear by user groups that typically protect their investments by lobbying for less restrictive regulations. Therefore, the desire to maintain access rights will be an important political determinant of the use of reserves in managing marine resources.

Potential On-Site Economic Benefits to Fisheries and Other Users

As Chapter 5 indicates, consensus is beginning to emerge about how some of the services produced from marine resources would change with MPAs and reserves. For instance, it is reasonably clear from much of the research to date that reserves will, under most circumstances, increase the biomass of exploited fish stocks, increase biodiversity, and allow recovery of the ecosystem to a more natural state *within* the reserves. These types of changes would produce important and valuable new services for direct, on-site, nonconsumptive and possibly some consumptive users. For example, it is likely that a more diverse and natural ecosystem would appeal to tourists and divers. Also, reserves would give fishery scientists and managers a baseline with which to compare undisturbed and exploited systems, which is especially valuable for increasing the accuracy of parameters used in fish stock assessment models. In addition, for people interested in the heritage values associated with protected and natural systems, reserves would also produce important new benefits. The total magnitude of these kinds of potential on-site benefits is an empirical question that has not yet been widely examined. Understanding the benefits within reserves from the protection and recovery of more natural systems will require further analysis of various kinds of ecosystem services that are not typically marketed. However, marine reserves may be the only method for preserving unique habitats and ecosystems.

Measuring Non-Market Benefits

Some of the services provided by marine ecosystems have market prices that can be adjusted to reflect their direct economic value. For example, the market prices of fishery products are commonly monitored and recorded in order to gauge the apparent values that consumers place on fishery products as well as

---

[1] *In re* Steuart Transportation Co., 495 F. Supp. 38, 40, E.D. Va. 1980.

the input costs used to provide these products. At the same time, market prices are not available for all services and, in some cases, may understate the true value of natural resource services. Market prices also may not give the correct "signals" about values that might be associated with either marine products or marine ecosystem services in the future. The challenge is to derive methods that can be used as market value "proxies" to assess an ecosystem's current nonmarket values where possible and to adapt those methods to predict what the values might be for future generations. Particular challenges include how to

1. assign economic values to on-site nonconsumptive services that benefit activities such as tourism, education, and scientific knowledge or to services provided off-site or indirectly through the site's role in the ecosystem;
2. incorporate externalities, such as damage to habitat or bycatch;
3. overcome technical difficulties in assessing the extent of the resource (i.e., marine biological diversity); and
4. account for what we don't know (complex ecosystem dynamics).

When market values are not available, proxy values have been computed to give at least a minimum (economic) estimate of how people value marine ecosystem services. Methods for measuring these proxy values include (1) hedonic values (Ridker and Henning, 1967; Brown and Mendelsohn, 1984; Garrod and Willis, 1993); (2) complementary marketed goods (Braden and Kolstad, 1991; Freeman, 1993; Hanley and Spash, 1993); and (3) surveys to determine values (often called the "contingent value" method) (Davis, 1963; Brookshire et al., 1976; Mitchell and Carson, 1989). The hedonic approach (HA) attempts to decompose the price of a marketed good into components that are associated with various attributes, some of which may be environmental. For example, one could gather data on property sales in an area that included some homes with beach front and decompose the sales prices into components that were associated with the dwelling characteristics, those that were associated with the value of bare land, and those that were associated with the aesthetic value of the ocean view. Similarly, analysts who study recreation values attempt to measure how attributes such as congestion, fishing quality, and other measures of environmental characteristics affect the amount people are willing to spend on the recreation experience (Bockstael et al., 1987). Both of these are "hedonic" techniques in the sense that they try to separate a single expression of monetary value into parts representing various characteristics of aesthetic and other valuable experiences (NRC, 1995).

The second method, using complementary marketed goods, is typically used in recreational valuation. Often, this approach uses a travel-cost model (TCM) to estimate the value of a particular site. Suppose, for example, that a particular lake is enhanced by restocking with native fish desired by anglers. Then a measure of the minimum of the individual economic values generated by this policy would be the increase in overall travel expenses incurred by people who come to this lake after it is restocked, relative to the number participating before the lake was improved.

The third method of deriving proxies for market values is the contingent value method (CVM). This consists of a survey method that places respondents in various hypothetical circumstances and asks questions about how much they might pay for an experience or how much compensation they would need to forgo an experience. For example, respondents may be asked whether they would be willing to pay a higher utility bill (a specified amount per month) to reduce electric power-related air pollution. This method is called contingent value because it elicits monetary valuation of hypothetical (or contingent) circumstances from the respondents.

The hedonic and complementary goods methods examine actual behavior and, hence, measure actual (revealed) willingness to pay for environmental goods. The CVM, on the other hand, measures individuals' hypothetical willingness to pay. It should be noted that contingent value studies may be used to measure willingness to accept the loss of some environmental services or opportunities or reduced quality. In theory, these should not be too different (see Willig, 1976), but in actual survey research they often are. All of these methods of developing proxy values for nonmarket services are based upon eliciting the current values of individuals participating today. An important issue, however, is whether these may understate the values that might be held by future generations (Krutilla, 1967).

An Example: Valuing Whales

How might these methods be used to value whales? A first step is to compare the different kinds of market values attached to whales. Whale meat is marketed in some countries; hence there is a market price based on whale consumption. At the same time, there are competing market values associated with the nonconsumptive use of whales.

For example, whereas whalers once set out from Lahaina, Nantucket, and other ports worldwide on multiyear voyages armed with harpoons, their descendants may set out on day trips from the same port, escorting passengers armed with cameras. Tourists are willing to pay significant sums for a whale-watching tour, mainly to experience whales in their natural environment. It is likely, in fact, that the market values of a whale-watching trip far exceed the market values associated with whale meat. Whales also have value through their ecological role in maintaining the natural abundance of other marine species, including commercially valuable fisheries.

Most people agree that whales are appreciated for more than simply their value as a marketed commodity such as meat or an object of guided tours. So a next step would be to try to compute the off-site nonconsumptive values that people place on whales, even when they have no direct contact or other physical interaction with them. These include the more difficult-to-measure existence values, bequest values, and heritage values that current generations derive from the simple knowledge that whales are part of functioning marine ecosystems. They are measurable only through survey elicitation methods such as contingent value surveys.

Putting a dollar value on these assets is contentious and technically difficult, but economists and others argue that attempting to compute some reasonable values is preferable to letting them be undervalued in a political process that often undervalues public goods (Hoagland et al., 1995). Importantly, in the cases where methods have been careful and sound, the values can be large. A real example of calculations quantifying nonmarket values involved asking how much the public felt deprived when the spill of oil by the Exxon Valdez polluted Alaska's scenically spectacular shoreline. In this well-known case, researchers surveyed households throughout the United States (excluding Alaska) and found that, on average, people were willing to pay about $30 to prevent another oil spill (Carson et al., 1992). The jury in the Exxon case awarded $5.3 billion in damages—a figure that was in the range determined by the contingent value studies. In principle, such dollar values could be determined for other marine ecosystems. Examples from other studies that use hedonic, travel-cost, or contingent valuation methods to estimate monetary values for marine and terrestrial reserves are presented in Table 4-2.

It should also be pointed out that assigning a monetary value to the existence of whales engenders a vigorous debate because some people consider such calculations irrelevant and possibly immoral—a misguided attempt to put all human values in economic terms. Just as profiting from slave labor is viewed as immoral, hunting an endangered species may be viewed as immoral by some, in part because of the deprivation extended to all future generations.

A last point is that the values expressed by current generations may not reflect the values that might be placed on certain environmental resources by our descendants. In fact, it is likely that as environmental resources become relatively scarce compared with manufactured goods, they will become more valuable. This places special responsibility on the shoulders of current generations to be precautionary when actions are irreversible.

So, although it is difficult to place fair market prices on these future values, they must nevertheless be incorporated into current political decisionmaking processes. Public trust resources, as a part of our cultural heritage, merit conservation measures such as marine reserves to prevent biological or functional extinction by current human activities.

From these examples, there are at least four categories of values that marine ecosystems might provide with implementation of MPAs:

1. Market values associated with consumptive uses, such as the value of

TABLE 4-2 Comparison of Monetary Valuation Estimates for Marine Reserves

Marine Area	Location	Year	Mean Value ($)	Model Type	Income (thousand)	Mean Age (yr.)	Mean Multiple Destination	Source
John Pennekamp, Key Largo	Florida	1988-1989	356-533	TCM	59	47	Yes	Leeworthy, 1991
Galapagos National Park	Galapagos	1986	439	HA	45	53	No	Edwards, 1991
Great Barrier Reef[1]	North Queensland	1985-1986	228^d 138^i	TCM	—	—	No	Hundloe, 1989
Martha's Vineyard[2]	Massachusetts	1989	164	CVM	109	52	No	Kaoru, 1993
Bonaire Marine Park[4]	Netherlands Antilles	1991	132	—	—	—	No	Dixon et al., 1993
Wellfleet Harbor[3]	Massachusetts	1994	66^d $87\text{-}111^i$	CVM	46^d 70^i	56^d 45^i	No^d Yes^i	Kaoru and Broadus, 1994
*Monteverde Cloud Forest[3]	Costa Rica	1991-1992	7^d 6^i	CVM	21^d 62^i	—	No	Echeverría et al., 1995
Nadgee Nature Reserve[3,5]	New South Wales	1979	3	CVM	50	36	No	Bennett, 1984

SOURCE: Hoagland et al., 1995.

1 = values are per person per year; *2* = annual willingness to pay for water quality improvements at three coastal ponds; *3* = annual payment for preservation in perpetuity; *4* = values are average economic impacts (gross revenues) per dive for 1991; *5* = estimated existence value only. Some analysts have estimated different values for domestic or residential (d) and international or tourist (i)

TCM: Travel Cost Model, HA: Hedonic Analysis; CVM: Contingent Valuation Method. See text for explanation.

NOTES: Table includes estimates from land-based and coastal reserves and illustrates some of the difficulties involved in making comparisons across sites. All values are per person per day, unless otherwise indicated, and are expressed in 1995 U.S. dollars using appropriate exchange rates and price indices. Where values were estimated as one-time payments (see *), an annual payment has been calculated by multiplying the one-time payment (capitalized asset) by a discount rate of 5%. In some cases, annual estimates may be similar to daily estimates if areas are visited only once each year. Care should be taken in making comparisons of these types because of differences in modeling approaches, socioeconomic characteristics, and choices faced by individuals surveyed for each area. The reader is urged to examine each of the relevant studies for greater detail.

increased or more stable fish landings in surrounding open areas, are well-defined and easy to measure. Market benefits that are more difficult to quantify, such as increased knowledge of marine organisms and ecosystems, can be assigned monetary values in terms of what people might pay for such knowledge or what indirect benefits society may derive from the knowledge. For example, many new bioproducts including pharmaceuticals have been derived from knowledge of marine organisms.

2. What the public may be willing to pay to experience the marine realm in a nonconsumptive manner is also a market value, measurable in theory if not always in practice. For example, divers place high value on experiences in marine habitats with abundant, diverse sea life, and these values may be higher as a result of establishing protected areas.

3. Even an individual who might rarely, if ever, visit the ocean may nonetheless place a monetary value on its biodiversity and its existence as a unique ecosystem.

4. Society also values species and habitats for their existence and for the knowledge that humanity is connected to nature in ways that reach back to the ancient past and that most hope will extend into the distant future.

These values may be expressed at different levels by various social groups and collectives. Sometimes, competing values will make it difficult to reach consensus on a policy for managing natural resources.

Costs of Marine Reserves and Protected Areas

What are the costs of marine reserves and protected areas? Or, to put it differently, which stakeholder groups perceive that restrictions in MPAs would cause them to suffer losses, relative to the status quo? As one example, marine-based tourism and recreation can grow out of an MPA designation and may tax local communities that do not have adequate infrastructure—or do not want to be inundated with tourists. Perhaps the most vocal and reluctant groups are commercial and recreational fishers who currently exploit areas that might be set aside as ecological reserves. Many fishers view the establishment of these reserves as a "takings" of their traditional fishing grounds, subject to compensation from the government, in the same light as landowners view eminent domain or other condemnation actions. In the view of many fishers, personal investments and life-style changes, as well as investments of time and money made by industries that service fisheries, have all been made with the assumption that fishers would continue to have unlimited access to marine habitats. Despite such beliefs, legal precedents on public trust resources suggest that the establishment of reserves would not be recognized as a taking, and fishery managers already use closed areas as one tool of fisheries management.

It is obvious that the acceptability of reserves is inversely related to their perceived costs to stakeholders. This raises an important question, namely, Are there circumstances in which the creation of reserves might be costless, or at least of low cost, to fishers? In fact, there may be such situations. The most likely candidates are circumstances in which a fishery has been dramatically overexploited. In fisheries for which past regulations and access rules have failed to protect the health of the resource, U.S. legislation *mandates* that rebuilding plans be developed (MSFCMA, 1996 [Section 303[a][10]]). When fishing capacity or effort is significantly larger than the stocks can support, it is clear that the remedy must involve reducing effort and fishing mortality to an acceptable level (perhaps zero) to allow the biomass to rebuild. Throughout the history of fisheries regulation, there have been some success stories in which significant cutbacks were made in effort, allowing stocks to recover. For example, the Pacific halibut is frequently cited for its remarkable recovery from the 1930 to 1960 and again in the early 1980s, as are the Alaskan salmon recovery during the 1970s and recovery of Atlantic striped bass in the 1990s. However, there are also stories of failure to enact necessary cutbacks, the most dramatic of which resulted in the collapse of the cod, Atlantic halibut, and groundfish fisheries off New England and eastern Canada.

The important point relevant to fishery reserves is that reducing catch to rebuild an overexploited fishery can be achieved either by uniformly reducing catch over the entire spatial expanse of a fishery or by geographically reducing catch by closing some areas and leaving other areas open. For example, given an even distribution of habitat and fish, a 50% reduction in exploitation rate from a particular fishery could be achieved by cutting the season length to less than half over the whole spatial expanse of the fishery or by completely closing a large fraction (50% or greater) of the spatial expanse for the whole season if the total fishing effort is constant and uniformly distributed. Models that compare conventional management methods with reserves suggest that under some circumstances, fishery reserves can provide maximum yields identical to conventional management. Hastings and Botsford (1999) developed a model (using several simplifying assumptions) suggesting that yields could be the same either by regulating exploitation of a constant fraction H of the whole stock or by setting aside a fraction c of the coast in a fishery reserve without additional regulation of fishing in the open area. Although the two approaches could be equivalent in terms of yield, the densities and catch rates experienced by fishers could be substantially lower under spatial management (Box 4-1). Yet how high are these costs compared to the benefits derived from reserves? Detailed information on spatial distributions of catch rates, strategies used by fishers, and effects of displacement of effort on ecological and socioeconomic variables would be needed to evaluate trade-offs between expected costs and benefits in specific situations. Another consideration would be the cost of enforcement. In the long-term, enforcement of a permanently closed area can be relatively efficient as the fishers become aware of the restricted area, with compliance monitored using a

BOX 4-1
Equivalence Between Fixed Exploitation Rate Strategies and Fishery Reserves?

A simple fishery model was used by Hasting and Botsford (1999) to compare yields under fixed exploitation rates (a popular conventional management strategy) and under spatial management. A population of resident adults with broadly distributed larvae was represented in the model. A key assumption was that all density dependence occurred when larvae settled and settlement rate depended only on the density of settling juveniles, not on the local density of adults. Under these conditions, recruitment is a function of total biomass of spawners, independently of how spawners are distributed in space. A result of this assumption is that identical maximum equilibrium yields could be obtained either by exploiting a fixed fraction H of the whole stock or by placing a fraction c of the coast in a reserve and exploiting all animals that recruit annually outside the reserve. Developing Hasting and Botsford's equations a bit further, it can be shown that the fractions H and 1 minus c that result in identical yields are related by the following equation:

$$1 - c = \frac{H}{1 - a(1 - H)} ,$$

where a is annual adult survival. Thus, when survival is zero, the fraction of coast open to fishing (1 minus c) is identical to the fixed exploitation fraction H under conventional management. For example, exploiting 20% of the stock over the whole area results in the same yield as closing 80% of the area and exploiting all the annual recruits outside the reserve. With higher survivals, larger fractions of the coast can be open to fishing to achieve the same equilibrium yields as obtained with any given exploitation rate H (Figure 4-1). If, for example, 80% of the adults survived from year to year, yields obtained under a 20% exploitation rate could be equivalently obtained by placing 45% of the grounds in a reserve and fishing all recruits outside every year.

Although the two management approaches could result in similar yields, the costs of extraction could be very different under the two regimes. Under spatial management, biomass outside the reserve would be exploited harder and therefore would be more depleted than when no restrictions are placed on fishing location. Fishers would thus experience much lower densities and catch rates when a substantial fraction of the biomass that sustains the stock is placed in a reserve. Using Hastings and Botsford (1999) model, the densities *prior to the fishing season* under the two regimes (for H and c resulting in identical yield) are given by

$$\frac{(\text{Density in open area})_{1-c}}{\text{Density under fixed exploitation rate})_H} = 1 - a(1 - H)$$

which may imply substantial reductions in catch rates, commensurate with the reductions in density (Figure 4-2). For the numerical example above, with 80% adult survival and $H = 20\%$, the densities under spatial management that would result in identical yields would be less than 40% of those under the fixed exploitation rate.

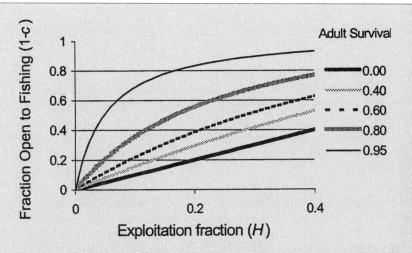

FIGURE 4-1 Exploitation fraction and fraction of the coast open to fishing that would result in identical equilibrium yields for different adult survival rates.

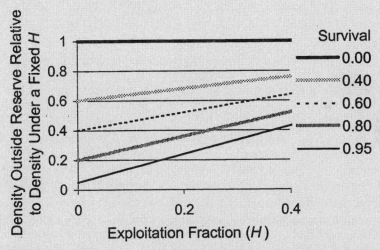

FIGURE 4-2 Density outside a reserve of area *c* at the start of the fishing season relative to density over the whole stock fished at a fixed exploitation rate *H*, when *H* and *c* are chosen to produce identical yields using the Hastings and Botsford (1999) model.

Reductions would be more severe once exploitation begins, because the rate of depletion under spatial management would be higher than under conventional, fixed-exploitation rate management. Although this model is oversimplified, it does illustrate some of the trade-offs that may occur in real situations.

vessel monitoring system (VMS). Table 4-3 presents some of the issues to be considered in evaluating potential costs and benefits of implementing fishery reserves. Clearly, standard management methods, such as effort controls (gear restrictions, days at sea, limited entry, etc.) and catch quotas, and spatial management methods (including fixed and rotating harvest zones, permanent reserves) can be complementary and often are used in combination.

Although there are circumstances under which marine reserves can benefit fisheries, there has not been much endorsement from the fishing industry. This may reflect the fact that reserve proposals are relatively new and hence not well understood by fishers. It also may reflect the absence of the fishing industry in the planning process. Other reasons for this resistance relate to the potential economic impacts of the reserve. First, altering the spatial extent of a fishery may change the relative advantages enjoyed by certain fishers, extending benefits to some while incurring costs to others within a fishery. Second, changing the spatial extent of a fishery will also tend to alter points of landing, distribution and processing channels, and the location of service providers that support fishing activities. For these fishers, fish dealers, processors, and industry suppliers, the real costs of spatial closures must be considered in implementing reserves so that closures do not disproportionately affect some individuals, companies, and communities relative to others in adjacent areas. Third, fishers realize that their rights of access are based on traditional and customary use, and are not supported by statute. The uncertainty associated with both resource levels and long-term access to the resources may produce an economic incentive to maximize profits in the short term at the expense of the long-term sustainability of the fishery.

Economic incentives need to be altered so that fishers make decisions that reflect the same long-term stewardship motives of managers and society at large (Hilborn et al., in press). This can be done either by stabilizing resource abundance or by making access to it more predictable and possibly guaranteed. Such convergence of motives would improve the implementation of fishery management measures, through either conventional approaches or reserves. Paradoxically, a major barrier to adopting reserves may be fishers' fears of permanently losing access to the fishery, even after the fish population has recovered.

Another issue behind the reluctance of fishers to consider spatial management on a par with conventional management—based on catch limits, time closures, and allowable gear—may be fear of a sort of double jeopardy in effort control. The displaced fishing activity might increase effort in the remaining fishing grounds to an extent that additional reductions in catch would be required to prevent overfishing. From the fishers' perspectives, this situation may appear to be "double jeopardy" in that they have given up half of their customary fishing areas and, subsequently, are forced to reduce their effort in the remaining open areas. Hence, achieving the cutback through a fishery-wide reduction in effort would be preferable to loss of part of the fishing ground with the potential for future reduction of effort in the remaining open area. This could motivate

TABLE 4-3 Potential Costs and Benefits of Fishery Reserves

Issue	Cost	Benefit
Yield	May lower catch, at least temporarily Uncertain benefits for mobile fish stocks For severely overfished stocks, large closures may be required with negative impacts on other fisheries	Longer-term stability of the resource Increased fecundity of resident stocks Increased future yields and recruitment Lower bycatch
Displacement	Hardship on local fishers and fishery-dependent businesses, including travel to more distant fishing grounds Increased impact on open areas	Reduces exploitation of overfished stocks (effort regulation comparable to conventional approach) Satisfies EFH provisions of MSFCMA
Enforcement	Requires enforcement of boundaries in addition to enforcement of fishery regulations in open areas. Costs of enforcement expected to increase with size of the area	Enforcement is relatively efficient, especially as fishers become aware of the boundaries and compliance increases as efficacy of reserves becomes apparent
Management	New requirements for monitoring and research—more data on stock life history, dispersal patterns, shift in effort, and habitat	Improved estimates of demographic parameters such as natural mortality rates for stock assessment Promotion of spatially explicit management to optimize allocation and use of marine areas and resources
Economic activity	Disproportionate impact on coastal community bordering the MPA Possible short-term reduction in profits	Higher potential for tourism and recreational opportunities Reduce potential conflicts with nonconsumptive users Increased future yields and recruitment Ensure against stock collapse from natural fluctuations or overfishing
Nonmarket values	Loss of customary fishing areas, and customary rights of access	Appreciation of more natural marine ecosystem Protection of scenic habitats and rare species

NOTES: The items listed under costs or benefits are possible, not definite outcomes of the implementation of fishery reserves. Attaching monetary values to these items, although possible, will require significant effort and research. EFH: essential fish habitat. MSFCMA: Magnuson-Stevens Fishery Conservation and Management Act.

fishers to lobby for fishery management that uses effort reduction instead of reserves, although effort reductions have other negative consequences for the fishery (see Chapter 3). More research is needed to understand how spatial and conventional effort controls compare in their effects on fishers' incomes, employment stability, and attitudes toward these different management approaches.

What about the impact of reserves on fisheries that are not overexploited? In these cases, it seems more difficult to argue that reserves are a win-win or cost-free policy for those whose livelihood depends on the fishery. Although reserves would certainly increase the flow of benefits associated with nonconsumptive uses of protected habitat to some constituents, these would seem to come at a direct cost to others who are being asked to give up exploitation opportunities in the reserves. Nevertheless, a reserve may be viewed as a beneficial investment under some circumstances even in a healthy fishery; following are some examples:

1. Reserves can be designed to enhance resource productivity, by protecting critical habitat, spawning, nursery, and juvenile grounds. Most fishers believe that critical habitat and young fish should be protected.

2. Reserves sited outside fishing grounds are easy candidates for protection but are less likely to benefit the fisheries because they are often population sinks (see Chapter 5).

3. Reserves may be seen as a complementary management method that provides more insurance against unforeseen fluctuations in fish populations and assessment uncertainty than can be achieved through conventional effort controls or catch quotas. Under certain circumstances, reserves could provide the same mean yield as conventional methods, but with greater protection against severe overexploitation or fluctuations of environmental conditions (Agardy, 2000).

ECONOMIC INCENTIVES

As described above, the benefits and costs of marine reserves are intricately bound up with perceived and de facto property rights to marine resources. Although property rights to marine systems are not secure in the same sense as titled land rights, access rights still have value to customary and potentially new users. Thus, the loss of access rights is viewed as a cost to customary users just as the creation of new access rights would be viewed as a gain. If marine resource rights were marketable, the rights would tend to flow to whoever valued them the most in market transaction (as demonstrated by individual fishing quotas; see NRC, 1999b). Usually, however, marine resource rights are not marketable; hence, disputes and conflicts over their allocation among various contenders tend to be resolved in the political arena, either in regional fishery management council meetings or through congressional action. The political process tends to protect entrenched interests, particularly when the rights involve services with clear economic value. If reserves are perceived to benefit one user group at the expense of another—for example, dive operators instead of aquarium fish suppliers—the losers will seek mitigation of the costs of their lost access rights. If the intent of a reserve is to benefit the fishery—for example, to use the reserve as a management tool to protect critical habitat for juveniles or to provide insurance against stock collapse—mitigation of the cost of lost fishing areas should not be as important an issue, since the same group (fishers) reap both the costs and the benefits of the action.

Several possibilities exist for compromise and conflict resolution in establishing reserves. First, reserves that have the lowest opportunity costs to current stakeholders might be chosen initially for development. These may be areas that have already been dramatically overfished since there will then be little to give up by setting them aside. Although fishers will expect to gain from emigration of fish from reserves (see Chapter 5), the issue to fishers is whether the spillovers are large enough to compensate for reduced opportunities from closures. Since the Magnuson-Stevens Fishery Conservation and Management Act (MS-FCMA) mandates that rebuilding plans be implemented, user groups will be required to incur the upfront costs associated with whatever methods are chosen to rebuild the fishery, and it is only a question of whether fishery reserves will be more effective than other management methods that uniformly reduce harvest throughout the area. However, even though overexploited fisheries appear to be logical choices in terms of the relative ease of implementing reserves, there are drawbacks to focusing only on overexploited fisheries. Some potential benefits of reserves may be realized only in stocks that are not overexploited, including protection of spawning stock, insurance against collapse, protection of habitat, and faster recovery after a disastrous event.

Second, where marine reserves are desired for their heritage values, but their establishment would cause specific user groups to bear substantial costs, it might be politically expedient to compensate fishers and other affected individuals for economic losses. This could be done with buyback programs or other capacity reduction methods that may have the auxiliary benefit of redirecting economic activity in fishing communities. Finally, conflicts between perceived winners and losers may simply play out in the political process. This option frequently delays the implementation of reserves, however, because the benefits appear diffuse while the costs are concentrated in a few, politically active industries. Even if reserves make sense from an overall national cost-benefit perspective, it can be difficult to overcome political barriers on the local level.

Property Rights and Rights-Based Management Methods

Most of the discussion thus far has presumed that management of the nation's fisheries will continue within the current system of imperfectly prescribed property rights in marine waters. In this system, the allocation of marine resources will continue to be the result of a political process of compromise among various stakeholder groups locked into conflict over the portfolio of marine ecosystem services. In addition, resources will continue to be exploited as common property, generally under conditions of either regulated open access or regulated restricted access.

Two very important consequences result from continuing either open- or restricted-access regimes. First, fisheries will continue to be exploited under the shortsighted approaches that develop among fishers with insecure rights to the resource, through continued pressure to overcapitalize, overexploit, tolerate bycatch, and produce lower-quality fish products of low value. These pressures arise because fishers do not have the proper incentives to act as long-term stewards of the resource. Most fishery economists have concluded that until the problem of insecure property rights is solved, commercial fisheries will continue to exhibit the historical symptoms leading to population collapses and broad economic and biological waste. Second, as discussed earlier, many of the public good benefits associated with marine systems tend to be underrepresented in political processes, with the result that the political system tends to undersupply services such as scientific knowledge, heritage values, existence values, and option values associated with environmental sustainability. Taken together, a critical issue in marine resource management is how to address the problem of common-pool resources to more effectively implement conservation measures, including marine reserves.

Changes are occurring in the United States and elsewhere in the world involving movement toward so-called rights-based management schemes (NRC, 1999b). These have various names, including individual transferable quotas and territorial use right fisheries (TURFs), but the important point is that they give fishers a guaranteed access right to a fraction of the total catch or area sanctioned for a specific fishery. Although these systems are not without controversy, they may generate a new stewardship ethic in the fisheries, and where carefully implemented, they have led to more effective management for long-term conservation goals.

Perhaps paradoxically, adopting rights-based management methods might make a system of marine reserves and protected areas easier to implement for the following reasons. First, as pointed out above, because the existing system of rights to marine resources is tenuous, various user groups lobby the political system to promote their own interests. This tends to favor commercial and direct user values over noncommercial and indirect public good values. Second, in a system that grants secure access to a given (sustainable) fraction of the resource, fishers could more confidently invest in future yields, such as those expected from a fishery reserve. Although both commercial and noncommercial users may stand to benefit from reserves, groups with a commercial interest may not be willing to risk short-term losses unless they are guaranteed a share of the long-term benefits. Without a system that adjudicates more secure access rights, it will be difficult to resolve the conflicts among stakeholder groups who can subvert an unpopular policy through "end-run" strategies such as lawsuits and political lobbying. This will delay or prevent reserve implementation and reduce the influence of proponents of reserves who are often underrepresented and over-shadowed by users with a more direct financial stake in the status quo (Box 4-2).

In addition to commercial fisheries, other sectors would be affected by reserves, some involving on-site extractive uses of the ocean's resources and others nonextractive uses. It is possible that systems of reserves might generate significant increases in on-site, nonextractive, uses such as tourism, recreation, diving, scientific research, and education, but the value of these nonextractive uses will have to be assessed and measures developed for valuing nonmarket activities such as scientific research and education. More cost-benefit analysis of these kinds of activities will be required to assess how they will be affected by implementation of marine reserves.

A final category of benefits, the existence or bequest values, also has to be evaluated carefully. These are perhaps the most difficult to assign monetary value for comparison with other costs and benefits. However, the available estimates obtained using the contingent value method have been significant (see Table 4-2). More studies of this type will be needed to complete a cost-benefit analysis of marine reserves.

Cost-Benefit Analysis of Marine Reserves Versus Conventional Management

The costs and benefits of marine reserves versus conventional management methods have not been thoroughly examined. There has been little experience with reserves and hence little empirical study of the costs and benefits of implementation. Also, most modeling studies to date have focused on either the biological or the economic performance of reserves, whereas more sophisticated integrated models are needed to facilitate comparison between reserves and conventional fishery management. Analysis of costs and benefits of reserves thus requires more research. Some general issues are discussed below that should be addressed in future studies.

For fisheries that are sustainable under conventional management, switching to marine reserves as the primary management approach will essentially substitute one effort control measure for another (see Box 4-1). From a cost-benefit standpoint, it is necessary to understand what would happen in the transition as fishers reallocate effort to remaining open areas. Reserves may turn out to be superior to conventional methods alone if there is a long-term gain in sustainable catch that exceeds the catch forgone from the reserve itself. However, the net economic profits from future catches must be discounted vis-à-vis any initial losses. Hence, for healthy fisheries, reserves may not offer dramatic benefits relative to catch and effort controls (reviewed in Milon, 2000).

Marine reserves also have been proposed as a supplement to conventional

BOX 4-2
Lessons Learned: Developing a Management Plan for the
Florida Keys National Marine Sanctuary

During the development of the management plan for the Florida Keys National Marine Sanctuary (FKNMS), opponents of the draft plan were successful in pressuring the National Oceanic and Atmospheric Administration (NOAA) to reduce the proposed area for ecological reserves in the sanctuary from 5% to 0.3%. NOAA coordinated the development of a comprehensive draft management plan that was released for public comment in March 1995. The public hearing process was extremely contentious. The most controversial aspect of the plan was the provision for "no-take" replenishment zones (renamed ecological reserves in the final plan). Release of the final management plan was delayed until September 1996, with most of the provisions for replenishment zones removed (Suman, 1998). Opposition to the plan was led by the Conch Coalition, an alliance of commercial fishers, treasure salvors, real estate interests, and other local residents, particularly those with valuable waterfront property. Subsequent surveys of commercial fishers, dive operators, and members of local environmental groups indicated that support for the reserves was high among dive operators (75%) and local environmental group members (76%) and low (24%) among the commercial fishers (Suman, 1998). Totaled across these three interest groups, 50% supported the implementation of reserves while 30% were opposed. There was a high level of alienation of commercial fishers from the public review process, many of whom (67%) felt that participation in the process did not matter and most of whom (60%) felt that the planning process had not been open and fair. The release of the detailed draft plan triggered the distrust of this key stakeholder group, who rejected the reserve concept despite NOAA's presentations on the fishery benefits of these zones (Suman, 1998).

Recent efforts to establish an ecological reserve at the Dry Tortugas, one of the original proposed sites in the FKNMS draft plan have been more successful. Stakeholders were involved in Tortugas 2000 from the outset through a working group comprised of 24 members representing commercial and recreational fishers, environmental groups, recreational divers, researchers, citizens-at-large, regional fishery management councils, and state and federal government agencies. The planning process outlined in Figure 4-3, led to a consensus agreement to create a 185 square nautical mile reserve. This alternative was approved by the Sanctuary Advisory Council in June 1999 and by the Gulf of Mexico Fishery Management Council in June 2000. Further approval must come from the National Park Service and the State of Florida, due to jurisdictional overlaps of the proposed reserve sites. The recreational fishing industry appears to be the most vocal in its objection to the Tortugas reserve proposal specifically (Florida Sportsman, August 2000) and to all closed areas in general.[a]

---

[a] http://www.asafishing.org/programs/govtaffairs/marineprotectedareas.htm and http://www.joincca.org/html/releases/2000/cca_takes_a_stand_against_no_fis htm.

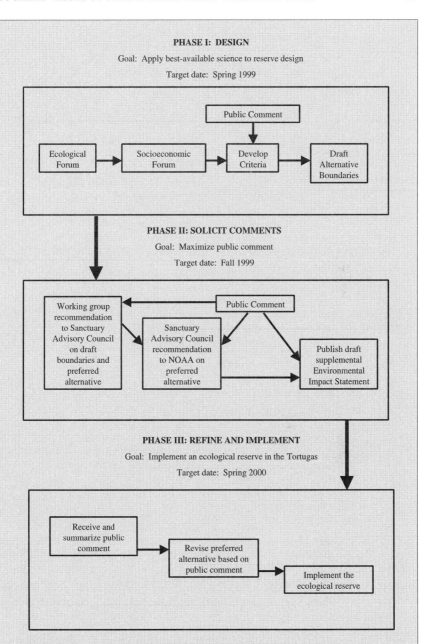

FIGURE 4-3 Tortugas 2000 planning process. SOURCE: http://fpac.fsu.edu/tortugas/images/process.gif.

management to provide insurance against uncertainty. As described in Chapter 3, conventional methods depend on stock assessments to set the appropriate effort controls or catch levels, but even under the best circumstances, stock assessments contain substantial uncertainty that may lead to overfishing. Marine reserves would provide a form of protection against uncertainty since the reserve would shield a fraction of the stock, which could help repopulate areas that become overfished. The insurance value of reserves has not been rigorously examined in a cost-benefit setting, and doing so will require an integrated bioeconomic analysis.

COMMUNITY INVOLVEMENT

Natural Resource Partnerships

Successful natural resource partnerships can be formed to implement MPAs if the potential partners can be identified and organized. Systematic social science research can contribute to identifying these partnerships by clarifying cultural differences in natural resource use, such as traditional ties of each group to marine ecosystems, knowledge, conservation ethics, and degree of compliance with laws and regulations. Just as many types of agencies potentially support and are involved in the planning of an MPA, so will there be many types of natural resource users who are potential partners in the MPA. The potential partners—including tourists, fishers, divers, marine product consumers, and conservation organization members—can be identified and assessed by systematic social science surveys and ethnography.

Much has been written about the advantages and disadvantages of having community-based management in which the interested public works with regulatory agencies and scientists to manage MPAs and other aspects of the marine environment (Pinkerton, 1989, 1994; Dyer and McGoodwin, 1994; Honneland, 1999). Community-based management, or co-management, refers to an alternative to top-down government regulation and strict market-based regulation models, in which collective solutions are sought for problems in managing common resources (McCay, 2000). It differs from top-down approaches that include stakeholder participation because it involves more than consultation—the community becomes involved in management of the resources. Community-based management of MPAs would involve any social collective or group that has well-documented connections to the marine ecosystem being considered for an MPA and would include clear leadership representation and some decisionmaking apparatus. The MSFCMA defines communities as geographic entities, although an argument could be made for broadening the definition to communities of interest or all stakeholders. Important communities of interest may exist at the local, regional, national, or international level.

Five main arguments exist for using a community-based management system for an MPA. Local residents and other community representatives should be involved because they

1. have some rights (formal or informal) in the involved coastal marine ecosystem;
2. have useful knowledge about the coastal marine ecosystem;
3. have personal or group resources that are needed for the MPA's operation;
4. will destroy or undermine the integrity of the MPA if they are not involved in its establishment and management; or
5. will support and enforce the rules of the MPA if they are involved in its establishment and management.

The issue of community management of marine ecosystems often dominates discussions about establishing MPAs. Most marine users perceive they have rights, however established, to features of the marine ecosystem that they use or to which they have become personally and culturally attached. The most common debate is over formal legal ownership and usufructory[2] rights derived from historic patterns of use. It is essential to identify and understand the role that such systems of formal and informal user rights have for MPAs. Formal studies of perceived rights should be conducted in order to help establish the core set of interests that should be represented on community-based management teams for an MPA.

Lay knowledge exists wherever people have used the marine ecosystem for any significant period, especially over many generations (Stoffle et al., 1994a). The validity and usefulness of such knowledge have not always been appreciated but must be addressed for managers to negotiate a common ground between scientists and users to implement policies with maximal effectiveness and minimal conflict. Community-based management programs must acknowledge that both lay knowledge and scientific research may have value (Stoffle et al., 1994b). Actions taken in the context of incomplete information require agreement among stakeholders, managers, and scientific researchers that regulatory actions are necessary and beneficial despite information gaps and conflicting perceptions of resource status.

Community-based marine partnerships often arise because there are insufficient resources to manage and enforce an MPA. Types of resources that communities provide range from physical facilities and equipment, to political support for establishing and operating the MPA, to day-to-day supervision and enforcement of MPA regulations (volunteer monitoring). At the national marine

---

[2] A usufructory right is the right to use something in which one has no property, that is, the right to take the fruits of property owned by another.

sanctuaries and the Dry Tortugas National Park, there is heavy reliance on voluntary support. Such arrangements are successful when the staff understand why communities provide resources and are trained to build and sustain these relationships.

One motivation for managers to employ community-based marine partnerships is fear of public backlash. This fear exists because restrictions placed on the customary users of marine reserve areas make managers vulnerable to public criticism and withdrawal of popular support. Nonetheless, a community-based partnership should not be designed to neutralize public criticism and increase public support without a genuine commitment to the process. Such actions are likely to be viewed as co-optive and will eventually be counterproductive. Instead, strong community-based partnerships should seek to develop mutual interests and mutual respect. Such partnerships generate confidence that all the parties will participate in management, share useful insights, and make the commitment to achieve the common goals for which an MPA was established.

In many nations, the development of MPA systems often has placed emphasis on biogeographic criteria and given socioeconomic factors less consideration, delaying participation until a late stage in the process, using a "sequential" approach. An alternative approach considers all of these factors at each stage of exploration, assessment, selection, and design. International experience overwhelmingly indicates that ignoring socioeconomic issues leads to failure of an MPA. According to Kelleher and Recchia (1998), two key lessons learned from establishing MPAs around the world are that (1) local people must be deeply involved from the earliest possible stage in any MPA for it to be successful, and (2) socioeconomic considerations usually determine the success or failure of MPAs.

The literature on MPAs is replete with examples of failure when the sequential approach is used. In recognition of this fact, the International Union for the Conservation of Nature and Natural Resources (IUCN, also known as the World Conservation Union) policy statement on MPAs made in 1988 includes the following provisions (Kelleher and Kenchington, 1992):

It is the policy of IUCN—The World Conservation Union—to foster marine conservation by encouraging governments, the non-governmental community and international agencies to cooperate in:

a) Implementing integrated management strategies to achieve the objectives of the World Conservation Strategy in the coastal and marine environment and in so doing to consider local resource needs as well as national and international conservation and development responsibilities in the protection of the marine environment;

b) Involving local people, non-governmental organizations, related industries and other interested parties in the development of these strategies and in the implementation of various marine conservation programmes.

It is also the policy of IUCN to recommend that, as an integral component of marine conservation and management, each national government should seek cooperative action between the public and all levels of government for development of a national system of marine protected areas.

This policy, which is based on decades of experience in all parts of the world, clearly indicates that IUCN members agree that socioeconomic issues have to be considered throughout the processes involved in identifying, selecting, and establishing MPAs.

Quoted below are some conclusions from the examination of a series of case studies of MPAs (Kelleher and Recchia, 1998). These case studies were from widely different geographic, social, and economic regions. Conclusions included the following:

- Socioeconomic considerations usually determine the success or failure of MPAs. In addition to biophysical factors, these considerations should be addressed from the outset in identifying sites for selecting and managing MPAs.
- Local people must be deeply involved from the earliest possible stage in any MPA that is to succeed. This involvement should extend to their receiving clearly identifiable benefits from the MPA.
- It is better to have an MPA that is not ideal in an ecological sense, but meets the primary objective, than to strive vainly to create the "perfect MPA."
- Design and management of MPAs must be both top-down and bottom-up.

Has the "sequential" approach been successful anywhere in establishing MPAs? Perhaps the best example of the sequential approach is that developed by the Canadian government, although many other countries or states have tried it in less systematic ways. As early as 1990, Canada had identified the 29 major biogeographic provinces of its marine environment and had developed an elegant systems approach to identifying priority areas for the establishment of representative MPAs. The outline of the method used is set out in IUCN's *Guidelines for Establishing Marine Protected Areas* (Kelleher and Kenchington, 1992). It exemplifies the sequential approach in that it was based on scientific considerations, without explicitly considering socioeconomic issues.

However, Canada's program to establish MPAs at the federal level has not been very successful. Since 1990, only one MPA has been established formally (the Saguenay-St. Lawrence Marine Park) and one tentatively (the Gully). This result can be compared with the experience of other countries, such as Indonesia, where socioeconomic issues are considered in parallel with ecological factors in an integrated way and where many MPAs have been established since 1990. However, the establishment of MPAs is only one measure of the effectiveness of the sequential versus the integrated approach. The next level of assessment is to determine whether MPAs have been effective in meeting their design goals.

Examination of MPA experiences worldwide led to the following conclusion (Kelleher and Kenchington, 1992):

> There is no simple or "turn-key" solution. What works for one nation or group of nations can rarely be transposed unmodified to another ecological or socio-economic environment. Nevertheless, there are strategic principles which are virtually universally applicable. One such principle is that a marine protected area is likely to be successful only if the local people are directly involved in its selection, establishment and management.

One almost universal aspect of human nature is people's suspicion of any action or program that may significantly affect their well-being if they have not been meaningfully involved in its design. If people as a group feel that they have not been part of the decision-making process, with genuine influence, it is usually difficult to obtain high levels of compliance from that group (Hanna, 1998). Instead, the group is likely to concentrate on the possible negative effects of the decision or action on its welfare. The best way to avoid losing support from one of these groups is to involve it in all aspects of a project. A person's strength of commitment to a course of action is likely to be proportional to the amount of "ownership" the person feels for that course. This sense of ownership is jeopardized by any exclusion from the decisionmaking process but is fostered when people can see that the plan considers their welfare in its design. Likewise, it is almost impossible in most modern societies to achieve long-term success in an action that affects the welfare of a local community if the community is opposed to the action. This has been demonstrated specifically in relation to MPAs (Salm and Clark, 1984, 2000; Kelleher and Recchia, 1998).

Experience from all parts of the world demonstrates that the apparent savings in time, human resources, and cost that might be achieved by excluding stakeholders—and thus avoiding conflict in early phases of a project—are illusory. When stakeholders are excluded initially, the later phases of a project often include conflicts arising from the reactions summarized above, which result in costs many times greater than the savings made through the initial exclusion of stakeholders.

5

Empirical and Modeling Studies of Marine Reserves

The goals for marine reserves, briefly described in Chapter 2, include conserving biodiversity, improving management of fisheries, and preserving and restoring habitat. These goals derive from societal wishes to preserve areas for the enjoyment of nature, to maintain functioning ecosystems, to establish replenishment zones for overexploited species, and to provide insurance against the uncertainty inherent in managing living natural resources. This chapter introduces the concepts underlying the use of reserves as a management strategy, describes empirical evidence from studies on existing reserves, and reviews various modeling studies.

CONCEPTUAL BASIS

Protect Intact Ecosystems

Current approaches to managing living marine resources typically address each species independently and ignore the spatial heterogeneity of marine systems. Consequently, policies may fail to protect some habitats and species. As a management tool, reserves have both disadvantages and advantages when applied to the diversity of marine species and habitats. The primary disadvantages are

1. establishing a reserve may result in the displacement of some fishers from customary fishing grounds with no impact on others, thus pitting the interests of one community against another;

2. closing an area may lead to increased human impacts on open areas;

3. to protect some species effectively, a reserve would have to be so large that it would be infeasible to implement and enforce; and

4. reserves, in general, will not be effective without continued conventional management of the area outside the reserves.

The primary advantages of using a spatial approach are

1. reserves should be relatively simple and inexpensive to enforce once the boundaries are established and recognized;

2. regulations may be tailored to specific habitats within the jurisdiction of regional management authorities (e.g., zoning seagrass beds as off-limits to destructive fishing gears);

3. reserves support conservation of the full range of marine resources, including habitat, biological diversity, and exploited species such as commercial fish stocks;

4. reserves provide unique sites for education and research on marine ecosystems, especially for comparison to areas altered by human activities; and

5. reserves provide "control" areas for determining natural mortality rates for different life-history stages, rates that are critical variables in stock assessment models (Box 5-1).

However, a marine reserve is envisioned to play a role in the ecosystem on a scale larger than its boundaries (Agardy, 1994). Reserves that are intended to fill heritage needs—for instance, to protect endangered species, collapsed habitat, or special features—could also provide protection for other vulnerable species that may support the recovery of areas disturbed by human activities. Examples include reserves for a habitat such as the Oculina Banks coral beds of eastern Florida (Koenig et al., 2000) or reserves intended to protect a specific geological feature such as the Texas Flower Gardens coral reefs (Gittings and Hickerson, 1998).

Preserve and Restore Habitat

Both biological diversity and productivity are fundamentally dependent on habitat, and loss of habitat is the leading cause of declining biodiversity (Wilcove et al., 1998; Wilcove and Wilson, 2000). In concept, reserves can protect and restore habitats that are critical for living marine resources. Structurally complex biological habitats often shelter breeding aggregations, provide nursery habitat, and supply food for adults (Ebeling and Hixon, 1991; Lindholm et al., 1999). Studies of areas in which the structural integrity of the habitat has been lost typically show a clear reduction in biomass and biodiversity (Dayton et al., 1995; Morton, 1996; Watling and Norse, 1998; Lindeman and Snyder, 1999; Koenig et al., 2000).

BOX 5-1
Case Study for Red Snapper

Management of the red snapper (*Lutjanus campechanus*) fishery is perhaps the most contentious issue in the Gulf of Mexico because it involves two valuable fisheries in the region. Red snapper stocks have failed to recover from overexploitation. Some have attributed this failure to bycatch of juveniles in shrimp trawls, rather than fishing by directed commercial and recreational fisheries (Goodyear, 1995; Schirripa and Legault, 1999). Red snapper fishers suggest that an essentially unfished stock of fish exists offshore, where it is unavailable to the fishery (Schirripa and Legault, 1999). They have argued that this presumed offshore stock provides sufficient spawning activity and that the onus for recovery rests on shrimpers who need to decrease bycatch of young red snapper. Juvenile bycatch represents as much as 70% of the entire fishing mortality of this species (Nichols et al., 1990; Nichols and Pellegrin, 1992; Schirripa and Legault, 1999) although it comprises only 1% of the total incidental take in the shrimp fishery.

However, the shrimp industry finds fault with National Marine Fisheries Service's (NMFS's) bycatch estimates and suggests that NMFS address alternative issues, including large-scale chronic (e.g., global warming) or acute (e.g., harmful algal blooms, hypoxia) events and the effects of the directed fishery on the age structure of the population.

A recent peer review of the red snapper stock assessment indicated that little effective management could be accomplished until basic information on red snapper population dynamics was acquired (Stokes, 1997). For example, the level of bycatch reduction required for stock recovery is uncertain without accurate estimates of natural juvenile mortality rates. It is essential to determine the age structure of the red snapper population in order to estimate natural mortality rates. Although a number of analyses of the red snapper data incorporate changing rates of natural mortality, the level of uncertainty surrounding these values is high (Goodyear, 1995, 1997; Schirripa and Legault, 1999), with some arguing for low levels, based on the red snapper's longevity (e.g., Camber, 1955; Wilson et al., 1994), and others for higher levels, based on the exclusion of regulatory discards from the samples (McAllister, 1997). The importance of bycatch reduction as a management strategy for red snapper recovery depends on the value used for the natural mortality rate (Stokes, 1997).

One way to obtain more accurate estimates of natural mortality would be to establish reserves and determine the age structure in these unfished areas. Such reserves also would promote red snapper recovery by protecting a fraction of the stock from fishing. Hence, reserves could play dual roles in promoting stock recovery and providing better estimates for population parameters used in stock assessment. This could serve as an additional incentive for the fishing industry to work on reserves with managers and scientists to make management more effective.

Reserves also can serve important functions in habitat restoration projects. Reserves act as controls (undisturbed sites) for evaluating the effectiveness of restoration projects at disturbed sites. Designation of areas as reserves could be used to protect sites so that restoration occurs naturally. There is a distinction between restoration of sites and mitigation that "creates" particular types of habitat where none existed previously. For example, artificial reefs placed on previously unstructured seabed and then protected from fisheries likely do not substitute for protecting natural ecosystems because functional biological communities also depend on the local oceanographic and environmental conditions and long-term interactions among species. Hence, artificial reefs may have little impact on overall productivity of reef fish populations and may in effect increase pressure on populations occurring on natural sites.

One category of habitat that has been severely reduced is coastal wetlands. Significant acreage of wetlands is still being lost each year to agricultural drainage, coastal development, oil spills, sewage, toxic chemicals, nonpoint source pollution, and destruction by introduced species. In California alone, nearly 90% of historical wetlands (once 3 million to 5 million acres, now estimated at 450,000 acres) are now used for agriculture and urban development, representing the greatest percentage loss in the nation. Thousands of acres of wetlands continue to be drained despite legislation (Food Security Act of 1985) that prohibits conversion of wetlands to agricultural use by landowners receiving federal farm payments. Recognition of the need to improve and restore estuarine habitat resulted in the creation of the National Estuarine Research Reserve System (NERRS) under the Coastal Zone Management Act (CZMA) of 1972. However, this program emphasizes developing and providing information to promote informed resource management rather than mandating specific protections (see Chapter 8).

The effects of habitat losses on fish stocks are diverse and ubiquitous, ranging from lower biodiversity in benthic habitats to reduced abundance of larger, economically important species (see Jones, 1992; Dayton et al., 1995; Kaiser, 1998; Pilskaln et al., 1998; Watling and Norse, 1998). Habitat loss from water diversion projects, hydropower dams, agricultural practices, and urbanization is a major factor contributing to declines in anadromous fish stocks (Rosenberg et al., 2000). Declines in shellfish and fish stocks may be caused directly by pollution or indirectly through algal blooms that either are toxic or deplete the water of oxygen (NRC, 2000a, b). More direct, fishing-related effects include destruction of biological structures (including oyster and coral reefs, sponge and bryozoan habitats) through abrasion and repetitive disturbance of mud-bottom communities by fishing trawls, dredges, or anchors, and loss of seagrass habitat from boating activities. Mitigating habitat loss will require water quality management and prohibition of activities that cause long-term declines in habitat quality and ecosystem health.

The movement toward an ecosystem approach to managing marine species gained momentum when habitat became a central theme of the Magnuson-Stevens Fishery Conservation and Management Act (MSFCMA) of 1996. This is embodied primarily in guidelines for applying the precautionary approach to management and recognizing the linkages between the goals of sustainable fishery production and conservation. The incorporation of concerns about essential fish habitat established the framework for an ecosystem approach to management, more akin to that developed for conserving terrestrial species (Zacharias et al., 1998).

Defining essential habitat is difficult because most marine and estuarine waters of the United States can be considered habitat essential to managed stocks and thus worthy of some level of protection. Structure and substrate in these habitats may vary from limestone outcroppings, to extensive seagrass beds, to kelp beds, to coral heads. Also important are mobile habitats, such as the extensive drifts of floating Sargassum in the Atlantic. Worldwide, many of these habitats have been declining for reasons ranging from nonpoint source pollution, to direct impact of fishing gear, to global climate change. In particular, loss of spawning habitat affects fish stocks in much the same way as fishing mortality of adult fish (Crowder et al., 2000). Hence, the dual mandates to adopt both precautionary and ecosystem approaches (FAO, 1995; NOAA, 1996a), coupled with that of defining habitat essential to sustain fisheries, make it imperative to evaluate habitat, determine its condition, and estimate its productivity. Life cycle patterns of the majority of economically important marine fish species can be reduced to several general habitat-related patterns (Koenig et al., 2000), some of which are quantifiable through development of abundance indices (Koenig and Coleman, 1998).

Establish Replenishment Zones

A marine reserve has the potential to play an important role in replenishing exploited marine ecosystems through the dispersal of larval or adult fish from the closed areas into regions where fishing is allowed (DeMartini, 1993). The idea that reserves will replenish fish stocks in open areas depleted by fishing is both promising (Carr and Reed, 1993; Quinn et al., 1993; Roberts, 1997a, Allison et al., 1998) and controversial (Coleman and Travis, in review). The promise lies in protecting fish so that they reach larger sizes, produce more offspring, and thereby increase the reproductive potential of a given species. One of the controversies, however, is that closed areas, in the absence of other measures to limit effort, may lead to increased fishing pressure operating outside reserve boundaries, such that the overall biomass of the stock decreases. In this circumstance, the anticipated benefits of the closure would not be realized, and there would be no net gain to the fishery.

Still, the idea of using closed areas as replenishment zones warrants investigation based on the biomass overflow and larval export hypotheses. The *biomass overflow (or spillover) hypothesis* suggests that the higher densities and greater average sizes of fish within a reserve will favor migration of adult fish into surrounding waters, thus augmenting or replenishing the population outside the reserve. Much of the empirical support for this hypothesis is based on anecdotal references. However, studies in the Philippines and Kenya documented increases in the populations of large adult fish in protected areas and subsequent population enhancement in adjoining regions (McClanahan and Kaunda-Arara, 1996; Russ and Alcala, 1996). Further evaluation in a variety of habitats with different species assemblages is required to determine whether spillover commonly provides significant replenishment for depleted fisheries.

The *larval export hypothesis* postulates that larvae will disperse out of the reserve and enhance recruitment in the fishing areas. Because reserves are associated with an increase in the density and size of fish and invertebrates, these populations should have higher spawning potentials and hence produce more eggs and larvae (Dugan and Davis, 1993; Bohnsack, 1996, 1998). If larvae disperse from the reserve into areas that provide appropriate nursery conditions, then stock enhancement in the open fishing areas is probable.

One of the better ways to replenish stocks and increase fishery yields is to protect stocks from growth overfishing. This is particularly difficult to do for species whose juveniles are caught incidentally in other fisheries. Designating nursery areas as reserves can protect juvenile fish from bycatch if the species are relatively sedentary during the juvenile stage. In the North Sea, large numbers of juvenile cod are killed annually in groundfish trawls (Horwood et al., 1998). Similarly, along the east coast of the United States, millions of dollars worth of yellowtail flounder (*Pleuronectes ferrugineus*) die each year as bycatch in trawl fisheries. In the Gulf of Mexico, nearly 70% of newly recruited juvenile red snapper are killed in shrimp trawls (Schirripa and Legault, 1999). If juveniles were protected in reserves, benefits would be delivered to the fishery when juveniles are exported from reserve areas to fishing grounds. Single-species closures used for plaice (*Pleuronectes platessa*) in the North Sea and mackerel (*Scomber scombrus*) in southwest England have resulted in increased yields by enhancing juvenile survival, even though neither area is completely closed (Horwood et al., 1998).

Provide Insurance Against Uncertainty

Insurance against the collapse of fish stocks is a primary, but elusive, goal of marine fishery management. However, managing fisheries and other living resources requires provisions for uncertainty in stock assessment (including limitations of the models and data used to estimate fish abundance) and for gaps in our understanding of the relative roles of environmental forces, fishing pressure, and management actions (see Chapter 3). Because uncertainty will always persist, fishery reserves have been proposed as a form of insurance to buffer potential failures of conventional management (Clark, 1996; Lauck, 1996; Lauck et al., 1998; Walters, 1998). In concept, protection of a substantial fraction of the fish stock in reserves will limit exploitation rates and prevent severe overfishing when other management measures fail (Bohnsack, 1998). In the event of collapse outside the reserve, a significant reproductive refuge would still be safeguarded that should serve as a source of replenishment to the depleted area and could accelerate recovery. Such a refuge would be effective regardless of assessment uncertainties, thus adding robustness to management, although reserves probably would not be sufficient without additional controls (Allison et al., 1998).

Does hedging against uncertainty provide sufficient justification to establish reserves? Is it likely that this form of insurance, with its own uncertainties, will be acceptable or preferable to conservative catch targets and thresholds within the context of conventional fishery management? One limitation of regulating fishing activity based on stock assessment is the inability to anticipate, and then compensate for, natural fluctuations in populations. A natural downward shift in abundance may not be detected until the closure of a fishing season, when it is too late to adjust catch levels for the actual size of the stock. The need for insurance also depends on the success of conventional management in preventing overfishing. Thus, the incentive for using reserves to ensure against stock collapse depends on the characteristics of the fishery and the quality of stock assessments, to the extent that these will determine the relative efficacy and implementation costs of spatial closures *versus* conventional management for ensuring sustainability. Highly mobile species, for instance, may require such huge spatial closures to achieve even modest levels of stock protection (Walters, 2000) that more conventional approaches would be preferable (e.g., improved assessments, reduced quotas, tightened effort controls). However, there are situations in which conventional management tools are particularly limiting. When assessments are poor or unreliable and not likely to improve, conventional management approaches may carry unacceptably high risks. In these circumstances, spatial closures may be the most practical alternative to provide insurance against stock collapse due to overfishing.

Fishery management is not the only source of uncertainty in complex, poorly understood marine ecosystems. For example, many estuaries, coastal ecosystems, and coral reefs are threatened by a variety of anthropogenic activities that include shipping, dredging, petroleum extraction, shoreline development, wetlands destruction, river and stream diversion, exotic species introductions, and pollution. These direct threats to habitats and ecosystems can indirectly affect fishery yields (Agardy, 2000).

Global climate change and interdecadal climate events such as El Niño, also place stress on coral reefs (Glynn, 1984, 1991; Glynn and D'Croz, 1990; Brown, 1997), further reducing the resiliency of reefs and other marine ecosystems to human impacts. It is now apparent that many marine organisms, especially those restricted to small geographic domains, face the threat of major declines or even extinction (Brander, 1981; Carlton, 1993; Casey and Myers, 1998; Roberts and Hawkins, 1999).

Will MPAs help protect against these stresses? In many cases marine zoning can reduce or offset the effects of shipping, dredging, and petroleum extraction, while more stringent regulation can reduce the effects of development and pollution. However, MPAs cannot protect an area from global warming or other large-scale climatic events (Allison et al., 1998), nor do they provide refuge from invasion by nonindigenous species (Simberloff, 2000). Although marine ecosystems are resilient and survive both short- and long-term climate variability as well as anthropogenic activities, resiliency has its natural limits (NMFS, 1999). Increasing human use, extraction, and modification of marine ecosystems have elevated the risk of long-term damage and provided the impetus to create MPAs as protection against unchecked expansion of human activities in marine ecosystems. In the wake of the catastrophic Santa Barbara oil spill in 1969, Congress passed the National Marine Sanctuary Act for "the purpose of preserving or restoring such areas for their conservation, recreational, ecological or esthetic values." Most commercial or recreational activities remain unrestricted in the sanctuaries, but the program establishes a framework for increasing the protection of vulnerable marine habitat.

Marine reserves could provide a mechanism for ensuring preservation of habitat areas of particular concern (HAPCs). The HAPC designation, developed by the National Marine Fisheries Service, refers to habitat judged to be particularly important to the long-term productivity of populations of managed species, to be especially vulnerable to degradation, or to represent examples of rare habitat types. Habitats with this designation include the Florida Middle Grounds in the Gulf of Mexico, which contains stony and octocoral habitat; the Oculina Banks in the South Atlantic, composed of thickets of fragile ivory tree coral *Oculina varicosa*; and a proposed HAPC in the gravel-cobble pavements in Closed Area II on Georges Bank in New England.

Distinguish Natural Variability from Human Impacts

Marine ecosystems are characterized by spatial and temporal variability that occurs at scales of centimeters to hundreds of kilometers and minutes to centuries. Global climate change and natural events that occur episodically over several years or decades, such as El Niño or other regime shifts, can result in long-lasting shifts in ecosystem composition and function. On a more local scale, the introduction of an invasive, nonnative species can affect an entire bay or large regions of coastline. Understanding the influence of human actions on marine systems is critical to evaluating the need for and effectiveness of management actions, but differentiating between natural and anthropogenic events is extremely difficult (NRC, 2000a). Any indicator of change in a system must be compared to a well-defined natural standard, or benchmark, against which the magnitude of the change can be evaluated to determine its cause and significance. Without control areas, such as reserves, that are relatively free from human influence to compare with areas altered by human activities, explaining the sources of variability becomes even more difficult. For example, much of what we know about fish populations is derived from fishery landings and is inherently biased, because estimates of fishing effort do not provide an index of fish abundance, and fishing changes the population dynamics of the target and associated species (NRC, 1998a). There is a significant need for fishery-independent sampling programs that include areas closed to fishing and other activities that disturb fish populations and habitat.

EMPIRICAL EVIDENCE AND INFERENCES FROM MODELS

Changes Within Marine Reserves

The finding that marine reserves improve species' abundance, biomass, and diversity came about almost by accident. Initial experience with reserves came from university research sites in Sumilon Island in the Philippines (Russ and Alcala, 1994), Leigh Marine Reserve in New Zealand (Walls, 1998), and Las Cruces in Chile (Castilla and Durán, 1985). In these research sites, fishing was prohibited to protect experiments from interference and to investigate effects of humans on coastal systems. However, exclusion also resulted in a dramatic transformation of the resident ecological communities. Careful studies at these reserves allowed researchers to gather quantitative evidence that closed areas could be used in marine conservation to restore or maintain a more natural ecosystem structure and to facilitate local recovery of depleted fish stocks (Russ, 1985; Ballantine, 1991; Castilla, 1999).

These initial studies indicated that establishing reserves had diverse effects, including rapid buildup in biomass (numbers and average body size) of previously exploited species and increased species richness (number of species per unit area)—that is, species appeared that had been absent prior to establishment of the reserve.

Effects on Abundance, Body Size, and Biomass

Halpern (in review) recently analyzed 76 studies of reserves covering a wide variety of locations, conditions, and enforcement levels. Most of these studies were conducted in newly formed reserves, so the effects represent early rather than long-term results. Because of the diverse taxa represented and the diverse methods of data presentation, Halpern derived a single aggregate measure to evaluate the results of each study. Many of the studies relied on "snapshot" comparisons of reserves with nearby unprotected sites, so the rate of change could not be determined. In any reserve, the composition and density of species in the area are expected to continue to change over time. However, in 69% of the reports, Halpern found overall increases in the densities of resident animals. Averaged across all of the studies, the density in reserves was almost twofold higher than in the unprotected sites. In general, local density appeared to increase with the length of time an area was closed (Table 5-1).

TABLE 5-1 Examples of Changes in Biomass and/or Density in Four Marine Reserves Closed for Different Lengths of Time

	Leigh Marine (New Zealand)	Apo Island (Philippines)	Saba Marine Park (Caribbean)	St. Lucia
Density	8	8	1.4	NA
Years closed	15	11	4	
Biomass	NA	8	4	3
Years closed		9	11	3
Species	Lobster	Predatory fish	5 exploited fish families	Haliotis kamtscharkana
Reference	MacDiarmid and Breen, 1992	Russ and Alcala, 1996	Polunin and Roberts, 1993 (Roberts unpublished data)	Wallace, 1999

NOTES: Density and biomass values represent increases relative to the density of comparable populations in neighboring areas open to fishing. NA = not available.

Ruckleshaus and Hays (1998), in analyzing similar pre- and post-closure changes in abundance in five reserves, found that reserves did not result in increased density of every species. Their analysis indicates that changes in community structure after the establishment of a reserve will not necessarily result in the increased density of all species affected by fishing.

In addition to abundance, the average size of individuals of a given species can change significantly within a reserve. In Halpern's aggregate analysis of reserve impacts, 88% of the studies found that the average body size of fish increased. Averaged across all target species and studies, the body size increased by about one-third (Halpern, in review).

Some long-lived species require decades to attain maximum size. Hence average body size within a reserve could continue to increase for many years following protection. Such increases have been observed after 15 years of protection in Egypt's Ras Mohammed Marine Park, where the average body weight of the lunar-tail grouper (*Variola louti*) increased three-fold in reserves compared to adjacent fished waters (Roberts and Polunin, 1993b). Similar findings were reported for northern abalone (*Haliotis kamtschatkana*) populations off Vancouver Island (Wallace, 1999); for quahog clams (*Mercenaria mercenaria*) within a closed area in Rhode Island (Rice et al., 1989); and generally for many fish species in reserves in New Zealand (McCormick and Choat, 1987) and the Mediterranean (Bell, 1983).

Biomass provides a more sensitive measure of reserve effects because it integrates changes in abundance and body size. For example, in and around Belize's Hol Chan Marine Reserve, fish densities increased in only a few families while fish biomass increased significantly after three years of closure (Polunin and Roberts, 1993; Carter and Sedberry, 1997). Biomass of economically important fish species increased two-fold in peripheral parts of the reserve and nine-fold in the center of the reserve compared to fished areas (Roberts and Polunin, 1994). Halpern (in review) found increases in 92% of the studies he reviewed, with the aggregated biomass in reserves on average two and a half times greater than in fished areas.

Effects on Species Diversity

Biodiversity encompasses all levels of organizational complexity, from genetic diversity to species diversity to ecosystem diversity (Chapin et al., 2000). Species diversity is the most commonly monitored aspect of biodiversity used in evaluating conservation areas. The concept of species diversity can be broken down into at least three components: (1) species richness, or the number of species per unit area; (2) species evenness, or the relative abundances of the range of species present; and (3) species assemblages, or the interactions between species (e.g., trophic structure). Changes in the assemblages or diversity of species can serve as key indicators of the impact of a reserve on ecosystem functioning.

In existing reserves, the point diversity of fish (i.e., the number of species per unit area of reef) generally increased when reserve sites were established in heavily exploited regions (e.g., Roberts and Polunin, 1993a; Russ and Alcala, 1996; Wantiez et al., 1997; Watson et al., 1997). Halpern (in review) found that 59% of the studies reported greater species richness in reserves. When averaged across studies, species richness had increased by one-third. In some cases, the apparent increased point diversity may have reflected an increase in abundance and hence an increase in the probability of observation, rather than the reappearance of a species in that area. Reserves may improve the long-term viability of some rare populations, especially for species that require relatively high densities for successful reproduction.

Expecting increases in all species is unrealistic given the complex linkages in ecological communities (Pimm, 1982), especially predator-prey relationships. Thus, species diversity may not be the best indicator of reserve performance. For instance, the density of some species may decline in an area closed to fishing (Ruckelhaus and Hays, 1998). Koenig et al. (2000) found that species diversity in the Oculina Reserve appeared lower when top-level predators were present than it did after removal of large predatory fish. In addition to real decreases in prey abundance from predation by species protected in the reserve, predator avoidance behavior may contribute to the apparent decrease. The effect of such avoidance behavior on assessments of species diversity will be particularly difficult to determine in deepwater habitats where census methods are performed remotely.

Establishing reserves in rocky intertidal habitats in Chile facilitated the recovery of the overexploited tunicate *Pyura chilensis* and the snail *Concholepas concholepas*. However, cascading trophic effects from the recovery of the snail resulted in lower overall diversity in this area (Davis, 1995; Castilla, 1999). These reports also documented the reduced abundance of mussels and herbivorous gastropods, and increased cover of algae or barnacles (Davis, 1995; Castilla, 1999). The Chilean studies illustrate how the trophic balance of an ecosystem shifts both with fishing and with subsequent protection, potentially benefiting some species at the expense of others.

Changes in species diversity associated with a closure will depend on the level of exploitation, life-history characteristics of the individual species, and potential for replenishment from surrounding areas. A common goal of reserves is to foster the recovery of species depleted directly or indirectly by intense fishing pressure (Fogarty et al., 2000). Species that are more vulnerable to exploitation and extirpation because of certain life-history characteristics (Table 5-2) are likely to benefit most from protection (Roberts and Hawkins, 1999). For example, the Looe Key National Marine Sanctuary in Florida supports six species of economically valuable fish that are absent from surrounding exploited

TABLE 5-2 Characteristics That Render Marine Species Vulnerable to Extirpation and Extinction

Characteristics	Vulnerability	
	High	Low
Population Turnover		
Longevity	Long	Short
Growth rate	Slow	Fast
Natural mortality rate	Low	High
Production biomass	Low	High
Reproduction		
Reproductive effort	Low	High
Reproductive frequency	Semelparity	Iteroparity
Age or size at sexual maturity	Old or large	Young or small
Sexual dimorphism	Large differences in size between sexes	Does not occur
Sex change	Occurs	Does not occur
Spawning	In aggregations at predictable locations	Not in aggregations
Allee effects[a] at reproduction	Strong	Weak
Capacity for recovery		
Regeneration from fragments	Does not occur	Occurs
Dispersal	Short distance	Long distance
Competitive ability	Poor	Good
Colonizing ability	Poor	Good
Adult mobility	Low	High
Recruitment by larval settlement	Irregular and/or low level	Frequent and intense
Allee effects[a] at settlement	Strong	Weak
Range and Distribution		
Horizontal distribution	Nearshore	Offshore
Vertical depth range	Narrow	Broad
Geographic range	Small	Large
Patchiness of population within range	High	Low
Habitat specificity	High	Low
Habitat vulnerability to destruction by people	High	Low
Commonness and/or Rarity	Rare	Abundant
Trophic Level	Rare	Low

[a] Allee effects occur when a reduction in population density has significant impacts on reproduction.

SOURCE: Roberts and Hawkins, 1999.

areas (Clark et al., 1989). In St. Lucia, some species are observed only in protected reserves (Roberts and Hawkins, 1997). However, there is no guarantee that a species will recover in a reserve if it has been eliminated regionally as a consequence of widespread fishing. The recovery of animals in a reserve in Jamaica's Discovery Bay is constrained by the virtual absence, to date, of recruitment by large snappers and groupers. In contrast, population rebounds have been observed for some species of smaller fishes that persisted in the area despite intensive exploitation for more than a century (Watson and Munro, in press). Hence, recovery in a reserve depends on the presence of source populations within the dispersal range of the depleted species and is especially critical for long-lived species (Roberts, in press). Recovery will be more rapid and assured when a reserve is established before exploitation essentially removes the entire breeding stock.

Contribution of Reserves to Replenishing Surrounding Areas

The contribution of reserves to replenishing fish stocks depends on the export of adults or young (larval or juvenile) recruits to fishing areas. Potentially, the more intense the exploitation outside the reserve, the greater is the density difference that will develop between protected and fished populations. The expectation is that the relative contribution of reserves to recruitment will be higher for intensively fished, non-migratory stocks. There are few studies demonstrating the replenishment of fish stocks on fishing grounds via export from reserves (Bustamante and Castilla, 1990; Tegner, 1993; Roberts, 1995). In some cases, this lack of data feeds the skepticism of the fishing community, but for some intensively fished species, models indicate that reserves have a high probability of increasing yields (see Attwood and Bennett, 1994; Murawski et al., 2000). An overview of studies of the effects of reserves on fisheries is presented in Table 5-3.

Replenishment from Spillover of Adult Fish

Direct confirmation of the export of fish from reserves to open fishing grounds is difficult and may require long time series to detect changes due to reserves over the normal fluctuations in the abundance of fish populations. A few mark-recapture studies show that significant numbers of fish tagged within reserves were caught in the adjacent fishing grounds and that fishing efficiency (catch per unit effort) increased outside the reserve boundaries (Attwood and Bennett, 1994; McClanahan and Kaunda-Arara, 1996; Johnson et al., 1999). Because such studies are so logistically difficult and costly, models show greater potential to improve our understanding of how marine reserves function in a regional context (see Gerber et al., in review). However, the conclusions drawn from a modeling effort are limited by the underlying assumptions on which the

TABLE 5-3 Overview of Fishery Effects in MPAs

Park	Source	Species Abundance	CPUE	Species Richness	Shannon-Weiner Diversity	Biomass	Size	Number of Fish	Density	Fishery	Design
Kisite Marine National Park, Kenya	27, 28	⇑		↔	↔		⇑	⇑			I/O, A
Cerbere-Banyuls, France	9	⇑					⇐			↔	I/O, A
Goat Island, New Zealand	8							↔	↔		I/O, A
Five Islands in New Caledonia	26			⇐		⇐	↔		⇑		B/A, I/O, A
Anse Chastanet	21					⇐					I/O, A
Apo Island, Philippines	1, 2, 24					⇐			⇐	⇐	Temp
Sumilon Island, Philippines	1, 2, 24					⇐			⇐	⇐	Temp
Maria Island, Tasmania	10						↔		↔		BACI
Medes Islands	11			⇐	↔		⇐				I/O, A
Okakari Point to Cape Rodney Marine Reserve, New Zealand	18							⇐			I/O, A

(continues)

TABLE 5-3 Continued

Park	Source	Species Abundance	CPUE	Species Richness	Shannon-Weiner Diversity	Biomass	Size	Number of Fish	Density	Fishery	Design
Barbados Marine Reserve	20						⇑	⇑			I/O, A
De Hoop Marine Reserve, South Africa	3, 4		⇑								I/O, A, Model
Tsitsikamma Coastal National Park, South Africa	5, 6										I/O, A
Okakari Point Marine Reserve, New Zealand	18							⇑			I/O, A
Soufriere Marine Management Area, St. Lucia	12									↔	Temp
Cousin Island Nature Reserve, Seychelles	14, 15	⇑				⇑			⇑		I/O, A, Grad
Sainte Anne Marine National Park, Seychelles	15	⇑				⇑					Grad
Baie Ternay Marine National Park, Seychelles	15	?[a]				?[a]			⇑		Grad
Curieuse Marine National Park, Seychelles	15	?[a]				?[a]			⇑		Grad

Site	Source						
Mobasa Marine National Park, Kenya	17		⇑		⇑	⇑	I/O, A
Saba Marine Park, Netherlands, Antilles	19, 23		⇑		⇑	⇑	Temp, A
Hol Chan Marine Reserve, Belize	19, 23				⇑	⇑	I/O, A
Ras Mohammed Marine Park, Egypt	22, 23		↔		↔	⇑	I/O, A
Exuma Cay Land and Sea Park, Bahamas	25		⇑		⇑	⇑	I/O, A [b]
Mayotte Island, Indian Ocean	16	⇑	⇑, ⇓		⇑	⇑	I/O, A
Carrie-le-Rouet, France	13	⇑		⇑		⇑	I/O, A
Looe Key National Marine Sanctuary, Florida	7				⇑	⇑	Temp

NOTES: *a* Unregulated reserve used as a reference site. *b* Only Nassau grouper (*Epinephelus striatus*), investigated. A: After reserve formation. B: Before reserve formation. BACI: Before/after control/impact design. I/O: Inside/outside reserve comparative study. Grad: Fishing intensity gradient design. Model: Model used to calculate fishery benefits. Temp: Temporal design, sampling before and after reserve formation. CPUE: Catch-per-unit-effort. ⇑ = increase, ⇓ = decrease, ↔ = no change.

SOURCE CODES: 1 = Alcala, 1988; 2 = Alcala and Russ, 1990; 3 = Bennett and Attwood, 1991; 4 = Bennett and Attwood, 1993; 5 = Buxton, 1993; 6 = Buxton and Smale, 1989; 7 = Clark et al., 1989; 8 = Cole et al., 1990; 9 = Dufour et al., 1995; 10 = Edgar and Barrett, 1997; 11 = Garcia-Rubies and Zabala, 1990; 12 = Goodridge et al., 1997; 13 = Harmelin et al., 1995; 14 = Jennings et al., 1995; 15 = Jennings et al., 1996; 16 = Letourneur, 1996; 17 = McClanahan and Kaunda-Arara, 1996; 18 = McCormick and Choat, 1987; 19 = Polunin and Roberts, 1993; 20 = Rakintin and Kramer, 1996; 21 = Roberts and Hawkins, 1997; 22 = Roberts and Polunin, 1993b; 23 = Roberts and Polunin, 1993a; 24 = Russ and Alcala, 1996; 25 = Sluka et al., 1997; 26 = Wantiez et al., 1997; 27 = Watson et al., 1996; 28 = Watson and Ormond, 1994.

model is based, particularly by assumptions about the degree of exchange among the reserves within a network or between reserves and the exploited habitats surrounding them (Palumbi, 2000).

Models have been developed to examine the potential contribution of dispersal from reserves to a fishery. For species with low adult mobility, individual size and population abundance should increase within the reserve, but with minimal spillover to adjacent areas. Indeed, Zeller and Russ (1998) found low rates of movement of the site-attached adult coral trout (*Plectropomus leopardus*) across reserve boundaries. At the other extreme, high rates of adult mobility between reserves and unprotected areas may drain the reserve population, especially when reserves are small. Fisheries for species with moderate rates of dispersal are predicted to benefit most from the establishment of reserves (Hastings and Botsford, 1999).

Effects of Reserves on Reproductive Capacity

The size of individuals and the abundance of target species frequently increase in reserves, yielding a higher reproductive capacity. Because fecundity (number of eggs produced) increases rapidly with weight in fish, the reproductive output of fish in protected areas should be significantly higher than that in areas where fishing removes a proportion of the larger, older fish. In actuality, fecundity increased two- to three-fold in two species of rockfish in a California reserve (Paddack, 1996) and six-fold for Nassau grouper (*Epinephelus striatus*) inside the Bahamas Exuma Cays reserve (Sluka et al., 1997).

In a simple modeling exercise, Bohnsack (1992) examined egg production by red snapper (*Lutjanus campechanus*) in the Gulf of Mexico with and without a 20% network of reserves. He estimated that closing 20% of the fishing grounds would increase egg production by approximately twelve-fold over the exploited stock (see Figure 5-1). The increase in reproduction derives from increased abundance of larger, older, more fecund females in the population.

Characteristics of Larval Dispersal

Most marine species exhibit considerable interannual recruitment variability (Dayton and Tegner, 1990). Dispersal distance depends upon time spent in the water column and oceanic transport (Hjort, 1914). The time spent in the water column depends on the time required for larvae to grow to settlement size, which is influenced by temperature and food availability (Bingham, 1992; Maloney et al., 1994). Hjort (1914) proposed the "critical period" for fish larvae as the time of first feeding coupled to successful dispersal to appropriate settling sites. It is during this time that fish must encounter prey in sufficiently high densities to ensure larval growth and survival. The longer they remain small, the longer they will be vulnerable to predation (Houde, 1987, 1997). In many cases, the survival

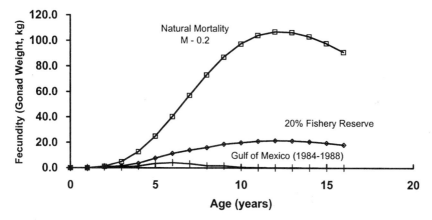

FIGURE 5-1   Estimated red snapper total fecundity (gonad weight) under conditions of natural mortality, average fishing mortality observed in the Gulf of Mexico between 1984 and 1988, and under conditions of a simulated reserve system that protects 20% of the population. The reserve system assumes that 20% of the cohort experiences only natural mortality, while the remaining 80% experiences the above fishing conditions. Used with permission from the author (Bohnsack, 1992).

of entire cohorts depends on weather and oceanographic variability that may either concentrate or disperse prey (Lasker, 1975), providing a biological explanation for the importance of the presettlement period (Jackson and Strathmann, 1981; Maloney et al., 1994). Recent results suggest that it is not possible to develop a generalized rule for larval dispersal or retention (Jones et al., 1999; Swearer et al., 1999; Cowen et al., 2000).

Many larvae, especially invertebrate larvae, require chemical cues to initiate settlement and metamorphosis. Suitable settlement sites that provide these cues are typically biogenic habitats—corals, algae, eelgrass, bryozoans—that also provide refugia and food for juveniles (Dayton et al., 2000). When these kinds of habitat are limiting, recruitment bottlenecks occur (Dayton et al., 1995; Lipcius et al., 1997; Bohnsack, 1998). Recruitment success in areas within and outside reserves will depend on the availability of suitable settlement sites.

Closed areas in the New England fishing grounds provide evidence of improved recruitment for some species. On Georges Bank and in the Gulf of Maine, more than 5000 nmi^2 were closed to bottom trawling and dredging in December 1994 in response to the critical decline of groundfish stocks. The intent was to improve recruitment by reducing bycatch of juveniles and preventing the disturbance of juvenile habitat in the closed areas. Collie et al. (1997) and Fogarty and Murawski (1998) offer excellent reviews of the history of direct and indirect impacts of fishing on Georges Bank where the cumulative impacts

TABLE 5-4   Dispersal Distances of Selected Marine Species

Genus and Species	Common Name	Average Dispersal (km)	Evidence	Reference
Echinometra meethaii	Pacific urchins	20	Genetics	Palumbi, in review
Solea vulgaris	Common sole	30	Genetics	Palumbi, in review
Riftia pachyptila	Tubeworm	20	Genetics	Palumbi, in review
Littorina cingulata	Gastropod	4	Genetics	Johnson and Black, 1998
Littorina littorea	Gastropod	20	Invasion rate	Carlton, 1982
Elminius	Barnacle	40	Invasion rate	Crisp et al., 1958
Carcinus mineas	Green crab	20	Invasion rate	Crisp et al., 1958
Cancer magister	Dungeness crab	50	Settlement pattern	Botsford et al., 1997

had been extremely damaging to demersal fish species and sea scallops (*Placopecten magellanicus*). They concluded that the three large closed areas aided recovery of yellowtail flounder (*Limanda ferrugineus*) and dramatically improved the abundance of large scallops (Collie et al., 1997; Fogarty and Murawski, 1998; Murawski et al., 2000). Scallop biomass tripled in the first 20 months after closure of the areas on Georges Bank (Anderson et al., 1999). Importantly, the current patterns and clustering of scallop fishing vessels at the downstream margins of the closed areas suggest that scallop larvae are exported and recruit to downstream areas (Collie et al., 1997; S.A. Murawski, National Marine Fisheries Service, personal communication, 1999).

Larval Dispersal Distances

Dispersal is facilitated for marine species having a planktonic larval stage. In fact, larvae of coastal species frequently appear in mid-ocean plankton samples (Scheltema, 1986). Most estimates of larval dispersal are obtained indirectly (Levin et al., 1993), derived from inferences about oceanography (Lee et al., 1994), larval biology (Emlet et al., 1987) or the genetics of adult populations (Palumbi, 2000; see Table 5-4). There are few direct observations of larval dispersal distances, except for species with low rates of dispersal (Olson, 1985; Stoner, 1992). Tegner (1992, 1993), after transplanting green abalone to a protected area depleted of adult stock, found larval recruitment significantly enhanced, with local recruitment observed from the transplants. This experiment terminated prematurely when the transplanted abalone were collected illegally.

Theoretically, species with relatively long larval periods (extending from a week to several months), experiencing current speeds of 0.1 m/s, could disperse more than 100 km downstream (Boehlert, 1996; Roberts, 1997b; Grantham et al., in review). However, recent reports suggest that average dispersal in some marine species is low and, hence, that long-distance dispersal may be less common at ecological time scales (Palumbi, 2000). For instance, species in many benthic habitats, such as sponges, several seaweeds, ascidians, and corals, have non-planktonic or demersal larvae and thus limited dispersal abilities. Also, long-term persistence of populations of marine species endemic to extremely small and isolated islands indicates that local retention must be highly significant (Hawkins et al., 2000). In simulation studies where larval movement is assumed to be passive, the results indicate that transport may be affected strongly by local eddies and current reversals (Lee et al., 1994; Limouzy-Paris et al., 1997), with limited dispersal from release points (McShane et al., 1988; Sammarco and Andrews, 1988). In his Caribbean study, Roberts (1996) concluded that surface currents might define dispersal patterns, with very large, often order-of-magnitude differences possible between the upstream and downstream settlement densities. Models of larval dispersal in reef habitats suggest that local retention on a natal reef (i.e., spawning site) is 10 times more likely than transport to downstream reef sites unless the spacing between reefs is about the same as reef diameter (Black, 1993). These studies indicate that knowledge of the dispersal patterns for larval replenishment will be important for effective reserve design and placement (see Chapter 6).

The issue of long-distance dispersal or local retention plays an important role in guiding decisions about the design of MPAs and reserves. Low transport rates for larvae of coral reef fish were determined using two methods: (1) micro-constituent analysis of otoliths in the U.S. Virgin Islands (Swearer et al., 1999), and (2) a mark-and-recapture tagging study involving millions of fish larvae around Lizard Island on the Great Barrier Reef (Jones et al., 1999). In both cases, a large proportion (roughly 50%) of larvae appeared to stay in local waters and recruit back to the home ranges of their parents. These studies are described in more detail in Chapter 7.

In other studies, local retention was rarer (e.g., Bertness and Gaines, 1993), but the fraction of larvae dispersing long distances was not determined. Larvae of species in protected areas appear to be more abundant just outside the protected area boundaries than at more remote sites, based on empirical and modeling results (Black, 1993; Palumbi, in review, 2000). If larval dispersal is low, even for species with planktonic larvae, then marine species, including some fishes, may require management on much finer spatial scales than previously assumed. Dispersal distances will affect the larval spillover distances to surrounding areas with implications for designing reserve networks that are self-sustaining (Figure 5-2).

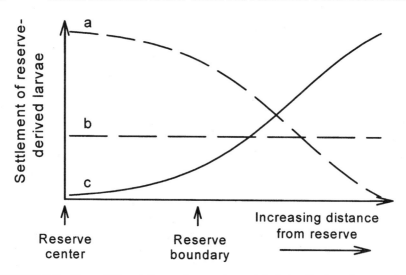

FIGURE 5-2   Three different dispersal scenarios and their implications for export of eggs and larvae from reserves: (a) short-distance dispersal leading to significant local retention and low dispersal away from reserve; (b) mixed strategy with both long- and short-distance dispersal leading to significant retention and export; (c) long-distance dispersal with low local retention and significant export. Reproduced from Roberts, 2000.

Genetic Analysis of Dispersal

In general, high dispersal potential is associated with high gene flow and hence low levels of genetic differentiation between populations over scales of hundreds to thousands of kilometers (Palumbi, 1992, 1994). Therefore, the level of genetic differentiation in adult populations can be used as an indication of the long-term dispersal history of a species. When a population is genetically distinct, this is an unambiguous indication of isolation in terms of recruitment from other populations. However, because a low level of gene flow is sufficient to maintain genetic homogeneity, the lack of genetically distinct populations does not necessarily correlate with high levels of interannual dispersal.

Johnson and Black (1998) documented an increase in genetic differentiation with geographic distance over a few hundred kilometers by measuring genetic relatedness in marine species over fine spatial scales. Palumbi (in review) used a similar approach to estimate average larval dispersal for a variety of marine species. These values ranged from 10 to 50 km and are similar to values estimated from the spread of invasive marine species (Crisp, 1958; Shanks and Grantham, in review) and from settlement patterns along coastlines (Botsford et al., 1997; see also Table 5-4).

Recruitment Sources and Sinks

The capacity of a protected area to replenish populations both inside and outside its boundaries may depend on whether the area will serve as a source or sink for new recruits (Allison et al., 1998). Sources produce "excess" recruits that spill over into surrounding areas, while sinks receive recruits but produce few of their own. In essence, a source is a productive habitat. Lipcius et al. (1997) have successfully measured sources and sinks for the spiny lobster (*Panulirus argus*) by measuring adult abundance, post-larval supply, juvenile density and nursery habitat. Evaluating these factors is critical for choosing a marine reserve site that will function as a source for the surroundings. Poor habitats are likely to be sinks, but often are suggested as potential sites for reserves because, typically, there will be fewer conflicts with users, especially fishers (Crowder et al., 2000; Dayton et al., 2000). However, locating a reserve in a sink area may increase fishing pressure on source populations and thus exacerbate population decline (Crowder et al., 2000).

The contribution of a reserve (located in a source area) to a fishery can be estimated conservatively as the fraction of the total biomass of the stock that resides in the reserve. There are several circumstances under which the reserve's contribution to recruitment is likely to be greater than this fraction:

1. *Where a fishery removes most fish before they have attained sexual maturity*: For example, in the fishery for red hind (*Epinephelus guttatus*) in Puerto Rico, 95% of the catch consists of immature fish (Sadovy, 1994). In this case, the reproductive potential within a reserve would exceed the potential in the surrounding fished areas. One way to estimate the reserve contribution to recruitment would be to calculate the reproductive biomass in reserves as a fraction of the total reproductive biomass in reserves plus fishing grounds.

2. *Where a species experiences strong Allee effects at spawning*: In this case, high population densities are necessary to trigger spawning and achieve fertilization. Allee effects are especially prevalent in sessile or sedentary invertebrates (Lillie, 1915; Levitan et al., 1992; Levitan, 1998; Dayton et al., 2000), such as scallops, clams, sea urchins and abalone. For such species, reserves may represent the only places where sufficiently high densities exist for successful spawning. In species subject to Allee effects, it is conceivable that a relatively small fraction of the biomass protected by the reserve could be responsible for the majority of reproduction (see Chapter 6, Figure 6-2).

3. *Where fishing reduces the abundance of one sex*: This is a particular concern for sequential hermaphrodites, like many grouper species, that change sex from female to male during their lifetime (see Box 2-1). Fishing skews sex ratios when it selectively removes larger fish that are predominantly male and thereby jeopardizes reproduction (Coleman et al., 1996, 1999), although there may be some compensation if sex change occurs at smaller sizes when males are

scarce (Roberts and Hawkins, unpublished data). The same pattern is found in many grouper species throughout the world (Shapiro, 1978; Nagelkerken, 1981; Bannerot, 1984; Beets and Friedlander, 1992, 1999; Buxton, 1993; McGovern et al., 1998). Potentially, fishery reserves could enhance recruitment by increasing male survival and sustaining a higher fraction of mature fish to breed.

Throughout the warm-temperate and tropical regions, many groupers, representing some of the more valuable food fish of these regions, undergo seasonal migrations to aggregate on traditional spawning sites. Fishers rely on this aggregation behavior to increase their catch rates by targeting fish during these vulnerable times (Johannes, 1978; Poizat and Baran, 1997). For example, the Nassau grouper (*Epinephelus striatus*) once had spawning aggregations that numbered in the tens of thousands of individuals but now have been drastically reduced (Colin, 1992; Sadovy, 1993; Sadovy and Eklund, 2000). The same problem afflicts dozens of other species, and one proposed solution is to place spawning aggregation sites in fully-protected reserves (Domeier and Colin, 1997; Johannes, 1998; Roberts, 1998b; Coleman et al., 1999).

Although protecting spawning sites would seem to be a sensible solution, Horwood et al. (1998) suggest that reserves in spawning areas could redirect fishing effort from spawning to nursery habitats and thus from mature to immature fish. Redirection of effort could then lead to more intensive fishing elsewhere, with possible habitat damage and increased bycatch of non-target species. Hence, redistribution of fishing effort is an important consideration in designing MPA networks. A combination of conventional management tools and reserves in spawning and nursery areas may be needed for some species.

Reducing Risk of Population Collapse

The concept of using reserves as buffers against overfishing has been explored to an extent using simulation models of fluctuating populations when catch rates are imperfectly controlled (e.g., Guénette and Pitcher, 1999; Sladek-Nowlis and Roberts, 1999; Crowder et al., 2000). Lauck et al. (1998), for instance, demonstrated that reserves could minimize risk by adjusting the percentage of closed fishing area relative to the catch rate in open areas. When the target catch rate in the open area greatly exceeded the level that maximized sustainable yields, a very large fraction of the area (greater than 60% in this example) had to be closed to maintain the stock at or above the level of maximum productivity. At lower catch rates, smaller closures achieved similar yields. Under the simplifying assumptions used in this model, the fishery reserve effectively protected a fraction of stock equal to the fraction of area closed.

In actuality, the level of protection conveyed by a reserve will depend critically on movement rates across the boundary (Guénette et al., 1998; Fogarty, 1999; Kramer and Chapman, 1999). When spillover of adult fish is high, more of the stock is exposed to fishing, and protection is lower. This "dispersal imbalance" (i.e., more fish leave the reserve than enter) will intensify if fishing effort concentrates near the reserve boundary (Walters, 2000). Simulation models often indicate that a large fraction (20–50%) of productive fishing areas may have to be designated as reserves to provide the desired level of insurance (Roberts, 2000).

Reserves for Migratory Species

When the mobility of adults is high, as in many pelagic and migratory fish species, reserves have often been discounted as an effective management tool. Whereas many coral reef fish have small territories as adults and may disperse during their planktonic larval stage, numerous fish species migrate hundreds or even thousands of kilometers annually (Harden Jones, 1968). Many high-value fish species, including cod (*Gadus morhua*) and herring (*Clupea harengus*) migrate long distances and so would obtain only intermittent protection from reserves as they pass through them. However, reserves on spawning grounds or in nursery areas for such species can offer protection and may be a management option. Even for highly migratory species such as swordfish (*Xiphias gladius*) or tunas, MPAs that protect nursery areas or vulnerable population bottlenecks may be effective as management tools.

Although it is generally assumed that all individuals of a migratory species migrate, recent evidence suggests that many individual fish cover relatively short distances. For example, most of 11,000 galjoen (*Dichitius capensis*) tagged in a marine reserve in South Africa remained within a few kilometers of their tagging sites, while about 18% dispersed tens to hundreds of kilometers (Attwood and Bennett, 1994). Hence, reserves may protect the less mobile individuals of these migratory species.

Effects on Habitat

Fishing also affects habitat and non-commercial species. However, most studies of reserve performance evaluate only a narrow range of taxa, focusing on fish assemblages. Studies of East African coral reefs demonstrate that fishing has broader impacts on the ecosystem, including changes in biogenic habitat. In this case, the fishery removed keystone predators of sea urchins, leading to urchin population explosions. As urchin grazing intensified, there was increased bioerosion of coral with reductions in coral cover (McClanahan and Shafir, 1990). In addition, sea urchins out-competed herbivorous fish for algae, leading to declines in the fishery for these herbivorous species (McClanahan et al., 1994).

In Jamaica, the ultimate cause of reef degradation following a mass mortality of sea urchins and two hurricanes was attributed to overfishing of herbivorous fish. After the removal of herbivorous fish, sea urchins were the primary players in controlling algal growth on the reefs. Following the die-off of the urchin population and damage to the reefs from two hurricanes, algal growth smothered the remaining coral (Hughes, 1994). A similar occurrence in St. Lucia confirmed that overfishing led to algal overgrowth on the reefs. In St. Lucia, shifts from coral to algal domination occurred only in fished areas, while in reserves, higher densities of grazing parrotfish controlled algal growth and prevented coral losses (Hughes, 1994). At a broader scale, coral cover in the Caribbean appears to be increasing only where there are well-managed marine parks and reserves (Ogden, 1997).

The above examples illustrate how reserves can prevent or reverse indirect fishing effects on habitat. Of greater concern in some regions are the direct and indirect effects of mobile fishing gears such as trawls and dredges (Safina, 1998a; Watling and Norse, 1998). This gear can destroy delicate biogenic habitats that may have taken centuries to develop (Dayton et al., 1995; Koslow, 1997). Most of the *Oculina* coral reefs of southeastern Florida have been reduced to rubble by trawling (Scanlon, 1998; Koenig et al., 2000). To avoid a similar fate, Norway established two reserves to protect deep-water *Lophelia* coral beds that recently became vulnerable after the introduction of "rock-hopper" trawls for fishing on rough seabeds. Also, there have been proposals to establish reserves on the Scotian Shelf off Nova Scotia to protect stands of deep-ocean soft corals that are vulnerable to damage from trawling gear (Kenchington et al., in press; Willison et al., in press).

Marine reserves clearly offer a reliable means to protect habitat, especially where fishing gear has been shown to destroy fragile, slow-growing, biogenic habitat such as corals. Often, there are no clear alternatives for protecting spawning sites and nursery grounds (Minns et al., 1996), although gear restrictions might be effective in some situations.

Research in marine reserves is now a fast-growing field of endeavor. Although they rarely appeared in the literature previously, the terms "marine reserves" and "marine protected areas" have increased dramatically in frequency in journal articles since 1993 (Conover et al., 2000). Studies evaluating reserve effects can be found in Roberts and Polunin (1991, 1993a); Dugan and Davis (1993); Rowley (1994); Bohnsack (1996); Allison et al. (1998); and Guénette et al. (1998). However, in the United States, there are very few closed areas that can be studied. The recent executive order from President Clinton (Appendix E), directing the Departments of Commerce and Interior to establish a national system of marine protected areas, could change this significantly.

6

Design

As with any enterprise, good design is fundamental for the success of marine protected areas (MPAs). This chapter evaluates how existing knowledge of marine ecosystems can be applied to the design of marine reserves and protected areas. Three important questions are covered: (1) How should the location of MPAs be chosen? (2) How large should MPAs be? (3) What kinds of zoning are useful in MPAs?

HOW SHOULD THE LOCATION OF MARINE PROTECTED AREAS AND RESERVES BE CHOSEN?

One of the more controversial issues in designing MPAs is deciding where to put them. While it may be possible to achieve consensus on the need for MPAs and agreement in principle on their size and the entities that should be protected, when it comes to choosing discrete sites, hostilities often break out. Frequently, it is the social aspects of locating reserves within MPAs that dominate arguments. For example, residents of an exclusive development hotly contested plans to include an ecological reserve adjacent to Key Largo in the newly created Florida Keys National Marine Sanctuary (U.S. DOC, 1994). They feared the reserve would prevent them from recreational fishing, or even landing fish, close to their homes. Commercial fishers responded similarly, suggesting that reserves would exclude them from their favored fishing spots. Clearly, social acceptance of the MPA plan, especially the location of reserves, is critical to

97

TABLE 6-1   Summary of Social and Economic Criteria Used to Select Marine Protected Area and Reserve Locations

Value Type	Criteria
Economic	Number of fishers dependent on the area
	Value for tourism
	Potential contribution of protection to enhance or maintain economic value
Social	Ease of access
	Maintenance of traditional fishing methods
	Presence of cultural artifacts or wrecks
	Heritage value
	Recreational value
	Educational value
	Aesthetic appeal
Scientific	Amount of previous scientific work
	Regularity of survey or monitoring work
	Presence of current research projects
	Educational value
Feasibility or Practicality	Social and political acceptability
	Accessibility for education and tourism
	Compatibility with existing uses
	Ease of management
	Enforceability

SOURCE: Adapted from Roberts et al., in review b.

successful implementation (see Chapter 4), but a balance between social concerns and biological function must be achieved. What methods are available for selecting functional reserves that meet these social and ecological criteria?

Kelleher (1999), building on previous work of the International Union for the Conservation of Nature and Natural Resources (IUCN, now the World Conservation Union), provided broad guidelines for selecting MPA sites, drawn from experience in the selection of terrestrial protected areas. He identified several classes of related criteria that bear on choice of a site: biogeographic and ecological criteria; naturalness; economic, social, and scientific importance; international or national significance; practicality or feasibility; and duality or replication. However, these guidelines neither offer guidance on how to prioritize these criteria nor provide advice on how to rank candidate sites according to each criterion.

This approach has been elaborated in recent papers (e.g., Salm and Price, 1995; Nilsson, 1998; Agardy, 1997). A summary of these criteria is provided in Table 6-1. All of these efforts focus on the problem of selecting individual marine reserves, but there is a growing awareness that this piecemeal approach to reserve establishment ultimately may fail to protect species and functional ecosystems. The implication derived from the broad dispersal capabilities and migratory behavior of many marine species, discussed earlier, is that even the largest reserves may fail to protect all resident species adequately (Ray, 1999). Widely dispersing or migratory species, for instance, will require networks of reserves. Site selection must take into account features at the scale of the reserve, but should also place the reserve into the larger context of the ecosystem within which it is embedded.

Ballantine (1997) proposed a series of principles for the development of regional reserve networks that build on some of the selection criteria for individual reserves. He argued that (1) all biogeographic regions and all habitats should be represented in reserves and (2) there should be replication of reserves within all regions and habitats among reserves. Ballantine also emphasized that within a network, reserves should have some level of connectivity; that is, they should be close enough for resident populations to interact through dispersal or migration. This connectivity among reserves, would be inherent in network design that fulfilled biogeographic, habitat representation, and replication criteria. In other words, fully representative networks with sufficient replication will inevitably contain reserves that are close enough together to interact effectively. Still, Ballantine (1997) offered little guidance as to how to weigh other criteria for selecting reserve sites.

Managers of marine resources would prefer an evaluation process that is more objective. Arbitrary choice of reserve sites could allow vulnerable areas to be degraded or destroyed, while other, more resilient habitats or species are protected. For instance, if criteria are clearly defined and agreed on prior to reserve selection, selected sites can be more easily justified. Much of the opposition that blocked zoning plans for the Florida Keys National Marine Sanctuary hinged on arguments that the locations of zones had not been chosen objectively. One of the original ecological reserve sites proposed, the Dry Tortugas, incorporated an area that satisfied neither conservationists nor fishers. Conservationists argued that it contained too little coral reef, while fishers objected to the extent of shrimp fishing ground that would be lost to them (Ogden, 1997). As a consequence, the siting of the Dry Tortugas reserve was deferred, and a more participatory design process was employed in planning the current site.

The development of more objective approaches could help meet the needs of managers for a more transparent process (Hockey and Branch, 1997; Roberts et al., in review a, b). The COMPARE (Criteria and Objectives for Marine Protected Area Evaluation) procedure described by Hockey and Branch (1997) marries the objectives of reserves to the selection criteria employed. COMPARE specifies 14 different objectives for reserves that are distributed among three categories: fishery management, biodiversity protection, and human use (Table 6-2). The authors propose 17 biological and social criteria that can be used selectively (not all are relevant to all objectives) to determine the value of a candidate site in meeting desired objectives. Sites are scored against each of the criteria, allowing straightforward judgments to be made of their relative value.

TABLE 6-2 Example of the COMPARE Procedure as Applied to Cape Point Nature Reserve in South Africa

	Objectives						
Criteria	Biogeography	Habitat Diversity	Rare or endemic species	Vulnerable Stages (all species)	Reduced Fishing Mortality	Vulnerable Stages (exploited species)	Adjacent Yield
Regionally representative	1						
Not conserved elsewhere	1		0			0	
High habitat diversity		1					
Includes fragile habitats		2					
Houses vulnerable species			1				
Protects rare or vulnerable stages			1	1	1	1	1
Pristine or restorable		2					
Special natural features	0						
Supports exploited species					1	1	1
Supplies adjacent areas						1	1
Large enough	1	1	1	1	0	1	1
Adjacent terrestrial reserve		1	1	1	2	2	
Aesthetically appealing		2					
Accessible to people							
Effective management	2	2	2	2	1	2	2
Satisfies social needs							
Preserves historical sites							
Totals for objectives	**5**	**11**	**6**	**5**	**4**	**8**	**5**
Percentages	**50**	**79**	**50**	**63**	**50**	**57**	**63**
Overall totals					**27**		
Overall percentages					**61**		

SOURCE: Hockey and Branch, 1997.

NOTES: 0 = ineffective; 1 = moderately effective; 2 = highly effective. Blank cells are inapplicable combinations. In this example, the reserve scores 61% for protection of biota, 55% for its contribution to fishery management, and 76% for provision of human uses, or 69% overall. This procedure has been applied to all of the existing reserves in South Africa and has helped evaluate the extent to which they are able to meet their objectives. It has also been valuable in determining what additional MPAs are needed to complete the country's national network.

Spawner Biomass	Research	Monitoring	Ecotourism	Low-Impact Recreation	Education	Exploitation	Totals for Criteria	Percentages
	2	1	1		2		7	70
0							1	13
	1	1	1		1		5	50
	2	1			2		7	88
	1	1	1		1		5	50
	1	1					4	50
	2	2	2		2		10	100
	0		0		1		1	13
1	2	2			2	2	12	75
						2	2	50
1	2	2	1	2	2	1	17	61
	2	2	1		2		14	78
			2	2	2		8	100
	2	2	2	2	2	2	12	100
2	2	2	1	2	2	2	26	93
			1	2	1		4	67
	0		0		1		1	17
4	19	16	13	10	23	7	136	
50	73	80	54	100	82	88	69	
21						88	136	
55						76	69	

Table 6-2 illustrates an example application of the procedure to an existing marine reserve in South Africa.

The COMPARE approach offers some welcome advantages over ad hoc site selection. It requires that sufficient and comparable data be collected for candidate sites. The simple, semi-quantitative evaluation method neatly summarizes the pros and cons of different sites and helps to pinpoint the deciding factors for making choices. The authors also suggest that management plans for reserves chosen using this approach can be guided by the objectives articulated during selection. It also goes beyond previous schemes in that the process can be used to build regionally representative networks of reserves. It is limited, however, because Hockey and Branch (1997) give only brief guidance on how to score sites according to each criterion.

Another approach that builds upon Hockey and Branch's (1997) efforts includes an evaluation scheme that aims to meld site selection more intimately with network development (Roberts et al., in review a, b). This approach departs from the COMPARE method, however, in that the criteria either are exclusively biological or are dependent on underlying biology. Their rationale is that if social and economic criteria override biological criteria, places of little biological value could be protected at the expense of areas with greater ecological value. The competing needs for biological relevance and social acceptance may be resolved by involving the stakeholder community at the outset in the process of identifying goals for the MPA and reviewing the criteria and data for site selection. Socioeconomic analysis of candidate sites is needed to identify and rule out areas that would be unacceptable to the community and, hence, impossible to implement or enforce.

Biogeographic and habitat representations are at the heart of most schemes for site selection. Leslie et al. (in review) used these criteria to design potential networks of fully protected reserves for the Florida Keys National Marine Sanctuary. They used computer-based selection algorithms to choose network designs that represented all habitats according to their relative coverage in the region. This exercise revealed that there are literally thousands of biologically adequate network designs. Selection from among these designs can be narrowed by identifying those network(s) with the lowest negative impacts on the surrounding communities.

In these examples, as well as other attempts to develop an objective approach to site selection, it is clear that the most subjective issue is the weighting of criteria. This reflects the reality that there is no formula that can be applied across the diversity of situations for planning MPAs. Therefore, involving stakeholders in every step of the process, from providing their knowledge of the environment and its resources, to making decisions about how to score sites relative to each criterion, is the most effective way to develop a cooperative, informed, MPA management plan.

If conservation of biodiversity is the goal, then ecological reserves must be located in places that will offer protection to the full spectrum of species and habitats. Box 6-1 lists one example of evaluation criteria, and Box 6-2 explains how these criteria could be applied. Biogeographic regions are usually defined on the basis of species composition and, in the marine realm, are often related to hydrographic features. Examples include currents (Emanuel et al., 1992), gradients of water quality (Roberts, 1991), patterns of productivity (Longhurst, 1998), and a complex mix of historical processes. These types of features have been used by Parks Canada for identifying representative marine areas in the development of Canada's National Marine Conservation Areas System. The Parks Canada criteria include

- geologic features (such as cliffs, beaches, and islands on the coast, and shoals, basins, troughs, and shelves on the seabed);
- marine features (tides, ice, water masses, currents, salinity, freshwater influences);
- marine and coastal habitats (wetlands, tidal flats, estuaries, high-current areas, protected areas, inshore and offshore areas, shallow water, and deep water areas);
- biology (plants, plankton, invertebrates, fish, seabirds, and marine mammals); and
- archaeological and historic features.[1]

Within biogeographic regions, candidate sites can be ranked on the basis of species richness or complementarity analyses used to ensure representation of species present (e.g., Roberts et al., in review b; Turpie et al., 2000). However, few sites have detailed data on species composition. In this case, habitats can be used as a convenient proxy for species. Ward et al. (1999) estimate that 93% of taxa will be represented in reserves that cover ≥40% of each habitat type. This is not to say that the presence of species of particular concern should not play a role in site choice. These species can be accounted for in the application of modifying criteria. When habitat representation is the priority, sites that have a high level of habitat diversity will receive a higher ranking.

In setting the goals for an MPA, the following questions should be addressed: Should different habitats be afforded different priority for protection? Are some more or less valuable or vulnerable? Should a greater proportion of some habitats and less of others (i.e., 5% of sandy shores and 30% of coral reefs) be protected? One approach is to protect habitats in proportion to their regional coverage (Ballantine, 1997; Roberts et al., in review a). Thus, if seagrass beds cover 25% of the total region, they would cover roughly 25% of the total area of reserves established. This does not mean that individual reserves would contain

---

[1] http://parkscanada.pch.gc.ca/nmca/nmca/program.htm.

BOX 6-1
Criteria for Selection of Ecological Reserve Sites
and Development of Networks
(Roberts et al., in review a)

All of the criteria have a biological basis or strongly affect species and habitats in candidate reserves.

* *Biogeographic representation.* All biogeographic regions should be represented and reserves should be replicated in each.
* *Habitat representation and heterogeneity.* All habitats should be represented and replicate habitats protected in different reserves within biogeographic regions.
* *Level of human threat.* Very high levels of human threat will exclude a site from consideration, but threats that can be mitigated could increase priority for protection.
* *Level of threat from natural catastrophes.* Sites that are foci for extreme natural disturbances should be avoided.
* *Size of site.* Candidate sites should be large enough to support viable habitats.
* *Connectivity.* Sites should interconnect with others through dispersal and migration.
* *Presence of vulnerable habitats.* Vulnerable habitats have higher priority for protection.
* *Presence of vulnerable life-history stages.* Vulnerable life-history stages, such as spawning sites, are afforded higher priority.
* *Presence of exploitable species.* Sites must be capable of supporting exploited species, even though the populations may be at very low levels at the time of implementation due to overfishing.
* *Presence of species or populations of special interest.* Endemic, relict, or globally rare species, for example, increase the value of a site, as would populations that are genetically distinct.
* *Ecosystem functioning and linkages.* Areas that link with and support other systems have a greater value than those that do not; similarly, sites that depend on links with other systems are vulnerable unless these places are also protected.
* *Provision of ecological services for people.* Services such as coastal protection or water purification add value to a site.

---

equivalent proportions of each habitat. Some reserves might consist entirely of one habitat, whereas others would contain a variety. Proportional coverage is the simplest argument to make, but there are other possible approaches. The answers depend on the needs of each region, but connectivity among reserves will likely require greater proportional protection of rare than of common habitats (Roberts et al., in review a).

In some cases, the primary objective for establishing an MPA may be to reduce or exclude pollution or mining. One of the main purposes of designating the Great Barrier Reef Marine Park in Australia was to exclude oil drilling and coral mining (Kelleher and Kenchington, 1982). Similarly, a primary motivation for establishing some of the National Marine Sanctuaries, such as the sanctuary in Monterey Bay, was the exclusion of oil and gas exploitation.

Potential sites may be scored against a series of six modifying criteria that affect their value as candidates for MPAs and reserves. Not all of these criteria are applicable to all objectives for MPA sites, but each is important for the development of successful networks of marine reserves:

1. Supplement or supplant conventional management of exploited species.
2. Protect rare species or vulnerable habitats. Vulnerable habitats generally include a biological component that is sensitive to disturbance and may take many years to regenerate.
3. Safeguard critical life-history stages, for example, spawning aggregation sites or juvenile nursery grounds (see Chapter 5).
4. Secure linkages among interdependent habitats. For example, communities present on rocky shores may derive their sustenance from adjacent kelp forests (Bustamante et al., 1995), and mangroves may protect offshore reefs from sediment carried by terrestrial runoff (Duke et al., 1997).
5. Maintain an ecosystem service such as the water filtration function performed by suspension-feeding invertebrates in bays and estuaries.
6. Provide connectivity among reserves for persistence of species or between reserves and unprotected areas for repopulation of exploited populations.

The application of these criteria depends on the goals of the MPA and zones designated as reserves. For example, the design of MPAs to improve fishery management may emphasize criteria 1, 3, and 6. Designing MPAs for conservation of biodiversity may emphasize criteria 2, 4, and 6.

It is logistically difficult to conduct experiments to evaluate connectivity. Roberts (1997a) suggested using prevailing current patterns to map connections among areas. However, others point out that currents may not reveal the true linkages if species actively control their pelagic dispersal (Swearer et al., 1999; Warner et al., 2000; Barber et al., in press). Biogeographic patterns of distribution provide a boundary within which connectivity may be evaluated. These boundaries may not depend on current patterns. For example, Barber et al. (in press) found that present-day current patterns could not explain patterns of genetic similarity among populations of mantis shrimps on Indonesian coral reefs. Generally, connectivity will be achieved automatically when networks of reserves are designed according to the other criteria, as found when these criteria were used to propose sites for marine reserves in Europe (Halfpenny and Roberts, in review). However, any obvious gaps should be avoided.

The schemes developed by Hockey and Branch (1997) and Roberts et al. (in review b), although applicable to the problems of reserve selection and network-

BOX 6-2
Decision Process for Developing Reserve Networks
(Roberts et al., in review b)

1. Define the goals of the network.
2. Define area of interest.
3. Divide it into possible reserve units—these may be defined in many ways, for example through grids of uniform-sized blocks (e.g., 10 km²), stretches of coastline, habitat classification schemes, or other means.
4. Select criteria for the evaluation of units that are appropriate to the goals.
5. Decide how to quantify the information needed to determine the level achieved for each criterion.
6. Assemble information on these units (e.g., species or habitats present, levels of threat).
7. Evaluation process:
   a. Characterize or score sites based on the following characteristics:
   Define biogeographic regions and then score sites based on what region they occur in. At this stage, sites could be stratified according to region and site selection decisions made separately for each region. The latter approach would be most useful where a large geographic area is being considered and there are many potential sites from which to choose.
   Define habitats within each biogeographic region for representation.
   Exclude sites that are subject to excessive levels of threat from human or natural sources.
   Include sites that are already reserves.
   Score potential reserves on the basis of habitat heterogeneity and representation criteria, ensuring that reserve units will be sufficiently large to include viable populations.
   Rank or score sites within each habitat type according to other modifying criteria.
   b. Set conservation targets for each of the criteria above (e.g., decide what proportion of the region and of each habitat to protect, what level of replication is required, levels of connectivity desired).
   c. Select among sites for inclusion in the network (this can be done with an algorithm or by a ranking or scoring method). Criteria may be given different weights at this stage in order to meet specific network objectives. Map the different biologically adequate networks of reserves that are possible.
   d. Ensure that alternative reserve networks resulting from the above selection process are sufficiently connected.
8. Use information on alternative, biologically adequate, reserve networks to inform final network selection according to socioeconomic criteria.

ing, may also be used to choose boundaries of individual MPAs or to designate zones for reserves within MPAs. Lafferty et al. (in review) show how these criteria can be applied to the problem of placing no-take zones in the Channel Islands National Marine Sanctuary in California.

The Importance of Developing Multifunctional MPAs and Reserve Networks

The fragmentation of objectives among different management entities often leads to a proliferation of uncoordinated zoning measures. McArdle's (1997) review of the 104 California MPAs, for instance, shows how the uncoordinated designation of sites results from overlapping, competing, and sometimes conflicting agendas of the different management agencies. This leads to a morass of legislation that perplexes users and may in the end harm, rather than help, conservation by giving the illusion of protection where little exists. For example, only 4 small MPAs, of the 104 in California, prohibit all recreational and commercial fishing.

Agencies need to cooperate in the establishment of MPA networks in order to reduce the costs of planning, implementation, and enforcement. Furthermore, implementation of a coordinated network of MPAs, using a selection scheme such as the ones described here, will ensure the greatest conservation benefit per unit area protected. The executive order issued by President Clinton on May 26, 2000, recognizes this need and directs the National Oceanic and Atmospheric Administration (NOAA), in cooperation with the Department of the Interior, "to establish a Marine Protected Area Center…to develop a framework for a national system of MPAs."

Biodiversity conservation, fishery production, and the full suite of ecosystem services depend on maintaining ecosystem integrity. This is a central objective for the creation of representative systems and fully interconnected networks (Agardy, 1997; Roberts et al., in review a, b). Fragmented initiatives are much less likely to safeguard ecosystem processes.

If reserves are to enhance fisheries through spillover of juveniles and adults, they must have "leaky" edges. As already discussed, larger reserves will have lower relative rates of export across their boundaries than small ones because the edges form smaller proportions of the area. For species that associate closely with particular habitats, reserve boundaries will be more porous (and spillover greater) if they are placed across areas of continuous habitat (Roberts, in press). Reserves whose boundaries are contiguous with habitat discontinuities will tend to have lower rates of spillover. However, reserves in which export rates are too high will fail on both fishery and conservation grounds. A balance between retention and spillover can be achieved at the reserve and network scales by incorporating a mix of boundary conditions.

As described in Chapter 5, sources are locations that supply recruits to other places, whereas sinks are places supported by recruitment from elsewhere without contributing to other areas (Pulliam, 1988). Crowder et al. (2000) modeled a system of sources and sinks for reef fish and found that at high fishing effort, placement of reserves in sink areas not only reduced the capacity of the reserve to support the fished population, but also concentrated fishing on source popula-

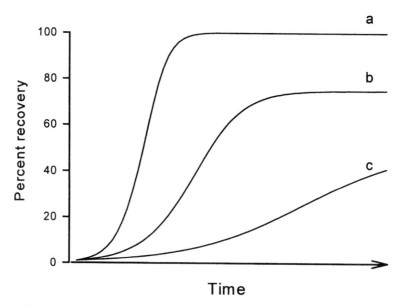

FIGURE 6-1   Three possible trajectories for community recovery (e.g., of biomass, abundance, or diversity) within marine reserves following protection from fishing at time 0: (a) high-quality habitat within reserve, (b) moderate-quality habitat, and (c) low-quality habitat within reserve. The rate and extent of recovery increase as habitat quality increases. Reproduced from Roberts, 2000.

tions. The model suggests that displacement of fishing effort to source populations could actually further the decline of a fish stock. Therefore, when reserves are established to benefit particular fish stocks the relative productivity of different areas should be considered. The higher the quality of the habitat, the greater are the expected extent and rate of recovery (Figure 6-1). It will not be possible to document source and sink sites for each individual species, particularly when a reserve is established for the full spectrum of biodiversity. For some species, source and sink areas may shift with fluctuating environmental conditions. A site that is a source for one species may be a sink for another, and a site that is a sink today may become a source in a few years. Establishment of a network of reserves can help address this complexity by covering a range of habitat types distributed throughout a region.

How Do International Political Boundaries Influence Reserve Selection?

Thus far, network development has been framed in the context of biogeographic regions, but biogeographic boundaries rarely coincide with political ones. How will this affect network development? If all countries were to commit to establishing comprehensive, representative networks of reserves, those networks should naturally link up across political borders. However, there are good arguments for greater collaboration among countries in planning reserves. For some countries, especially those with short coastlines, local initiatives may fail to produce the desired benefits unless similar initiatives are implemented by adjoining countries with which there are strong resource linkages.

In politically diverse regions, such as the Caribbean, the number of possible partner nations with which countries have to coordinate marine resource management seems daunting. However, maps of current patterns in the Caribbean reveal approximate upper bounds to dispersal distances for pelagic larvae with different larval durations (Roberts, 1997a). Potential dispersal routes and distances can be used to identify upstream and downstream partner countries. For example, recruitment on coral reefs in Florida probably includes inputs of larvae from reefs off Cuba and several Central American countries; therefore the quality of resource management in these countries should concern managers in Florida (Ogden, 1997).

Cross-border reserves illustrate a second way in which international collaboration can be beneficial. Border areas are frequently regions of political tension, with conflicts arising over vessels from one nation fishing in the waters of another. Fully protected reserves that straddle borders could help reduce conflict and, at the same time, improve the condition of marine resources in both countries. Such reserves have been proposed for the Hague Line, which defines the border between eastern Canada and U.S. waters in the Gulf of Maine (McGarvey and Willison, 1995; http://www.atlantisforce.org, June 2000), and for the waters separating the U.S. Virgin Islands and the British Virgin Islands. There is already a trinational reserve project under way in the Caribbean, encompassing the waters of Belize, Guatemala, and Honduras (Heyman and Kjerfve, 1999). This project is underpinned by knowledge of current patterns that link the resources of all three partners. Networks of MPAs and reserves that are designed to provide regional benefits across national boundaries can distribute the opportunity costs of establishing protected areas among coastal communities and countries.

Such international reserves could also help resolve territorial disputes. McManus and Meñez (1997) have highlighted the potential for an international reserve in the Spratly Islands of the South China Sea. Coral reefs of these islands are among the least impacted in Southeast Asia, and they may be a source of larvae for reefs in countries bordering the region.

Can Reserves Contribute to Conservation and Management of the Open Ocean?

The overwhelming majority of reserves established to date have been in coastal areas, extending at most to the limit of the continental shelf where the impact from human activities is most obvious. These places are intensively used by people and receive the most direct input of pollutants from land. However, open ocean areas are also impacted by human activities, raising the question of whether reserves should be established in the high seas.

Human impacts on the deep sea are increasing, with dumping, fishing, and mining the three most significant threats to deepwater habitats. Deep-sea fish species (fish occurring at depths of 200 to 2,000 m) have increasingly become the target of commercial fisheries, as trawling effort has shifted from continental shelves to deep slopes and seamounts (McAllister et al., 1999). Fishers regularly trawl at depths up to 2 km, with about 40% of the world's trawling grounds now located in deep water. Trawling for orange roughy (*Hoplostethus atlanticus*) off Australia and New Zealand has stripped seamounts of their unique invertebrate fauna (Dayton et al., 1995). Recent research has found high levels of endemism on these seamounts, raising the possibility that undescribed species are becoming extinct as a consequence of damage from trawling (de Forges et al., 2000). Reserves could provide these deep-sea communities much needed protection from collateral damage due to fishing as well as damage from mining or dumping. It is notable that the collapse of fisheries in shallower waters has often stimulated governments to promote deepwater fisheries (Haedrich, 1995; Moore, 1999).

Discovered less than 25 years ago, hydrothermal vent ecosystems have become an important topic of study in marine science and now are emerging as a focus for conservation as well. The fauna at the vent ecosystems occurs nowhere else in the deep sea, and a high level of endemism appears to exist at many vent sites. The geological and geochemical characteristics of the hydrothermal vents are also unique, and the polymetallic sulfide deposits formed at spreading centers are a potential source of economically valuable minerals such as copper ores.

Creation of marine reserves at deep-sea hydrothermal vent sites has been proposed for two quite distinct reasons (see http://triton.ori.u-tokyo.ac.jp/~intridge/reserve.htm). First, in view of proposals to mine the mineral wealth found at hydrothermal vent sites, reserves have been proposed as a means of protecting some fraction of these unique ecosystems from destruction. Biodiversity concerns loom large in view of the endemism associated with the vents. Second, as research programs have developed, conflicts have arisen between "observational" studies to provide long-term data on the development of vent ecosystems and "manipulative" studies to collect biological or geological specimens. Observational studies require closure of some areas within the vent ecosystem for ecological reserves that may be monitored, but not altered, by human activities.

Scientists working at vent sites are encouraged to communicate their needs through a Web site (http://triton.ori.u-tokyo.ac.jp/~intridge/reserve.htm) established by the InterRidge program, an international consortium of vent scientists. This Web site provides a single source of information on proposed vent studies, enabling scientists to anticipate and resolve potential conflicts between research programs. Formal establishment of reserves at hydrothermal vent sites by government agencies has commenced. In 1998, the Canadian government established the first pilot MPA at a hydrothermal vent site in the northeastern Pacific, the Endeavor Hot Vents Area, which is part of the Juan de Fuca Ridge System (see http://www.oceansconservation.com/mpa/related/fsendeav.htm).

The open ocean, like deep-sea regions, has seldom been considered as a potential location for reserves. Fisheries of the open ocean are heavily exploited, currently for migratory species such as tuna and swordfish, but formerly for whales. Over time, the scale and efficiency of exploitation have increased relentlessly, leading to significant declines and overfishing in many of these fisheries (Safina, 1998b). Impacts of exploitation also extend far beyond target species (Dayton et al., 1995). Tens of thousands of marine mammals and birds are caught annually as bycatch in drift nets, and thousands of turtles and birds are killed by longlines. Swordfish fisheries kill several sharks for every swordfish landed. There is an urgent need to protect species of the open ocean, and MPAs may be an important tool to achieve this objective.

A problem that complicates efforts to manage deep- and high-seas fisheries is that many operate in international waters. Nations are less inclined to regulate activities on the high seas than in nearshore waters because they lack jurisdiction over such regions. However, there appear to be many circumstances under which MPAs might provide benefits in these regions. There are some precedents, including the Indian Ocean Whale Sanctuary, the Antarctic Treaty, and the Torres Strait Treaty between Australia and Papua New Guinea. Enforcing offshore MPAs will be a challenge, but the development of satellite tracking technology for fishing vessels (vessel monitoring systems) and other technologies may solve these problems.

HOW LARGE SHOULD MARINE PROTECTED AREAS BE?

The question of size has two elements: (1) How much of the sea should be protected in total? (2) How large should individual reserves be?

How Much of the Sea Should Be Protected?

The question of how much of the sea should be protected from human disturbance is one of the most vexing issues surrounding marine reserves. In the terrestrial realm, the World Conservation Union (IUCN) has recommended that 10% of each country's land area be set aside in protected areas. However, many think that the more open character of marine ecosystems requires that higher targets be set, with 20% most often cited as the appropriate range (Schmidt, 1997). The U.S. Coral Reef Task Force (USCRTF, 2000) has recommended that 20% of coral reefs and associated habitat types receive protection in reserves. Although the 20% figure is widely quoted, it is often criticized as being arbitrary and unscientific. To many fishers, it seems unnecessarily high. When the Reef Fishery Plan Development Team (RFPDT, 1990) recommended the protection of 20% of the continental shelf off the southeastern United States in 1990, its report was met with incredulity and hostility from the fishing industry, and the proposal was shelved.

This section provides an overview and synthesis of studies examining the question of how much area in the sea should be protected. Table 6-3 summarizes 35 studies that approach this question from a range of perspectives, and a description of each study is provided in Appendix G. The table is organized according to the principal issues addressed in each study: (1) ethics, (2) risk reduction, (3) yield maximization, (4) preservation of biodiversity, and (5) increasing connectivity among reserves. Many of the more influential studies are discussed in detail in the text.

The goal of protecting 20% of the sea was first proposed by the Reef Fishery Plan Development Team (RFPDT, 1990). The rationale come from a fishery model indicating that recruitment overfishing could be avoided by maintaining stocks at or above 20% of their unfished biomass (Goodyear, 1993) and from experience with habitat closures for several invertebrate species in the southeastern United States (Bohnsack, 1996). Later analyses suggested that some stocks should be kept above 35% or even 40% of their unexploited biomass (Mace and Sissenwine, 1993; Mace, 1994), but the 20% figure for reserves persists. Others have suggested that even higher targets are necessary, such as maintaining 60% or even 75% of unexploited biomass if reserves are used as the primary management approach (Hannesson, 1998; Lauck et al., 1998; Mangel, 2000).

A common argument for using reserves in fishery management is to provide insurance to counter the uncertainty inherent in conventional management (Chapters 3 and 5). Lauck et al. (1998) showed that irreducible uncertainties in estimates of population size and fishing mortality make it difficult for managers to avoid driving stocks below critical target levels. Large closures provide a risk-averse strategy for meeting management objectives. Models suggest that reserves covering between 30% and 60% of management areas would offer risk reduction (Table 6-3).

Several common conclusions can be drawn from a broad range of biological and economic models that address the role of reserves in improving fishery yields. First, reserves are likely to support increased yields for overexploited fisheries, but considerable areas must be protected to achieve such benefits. Second, as fishing pressure outside reserves increases, the size of the area in reserves must also increase to sustain the population. Third, without other management measures, highly mobile and migratory species will require very large (70-80%) closures. However, most proposals for establishing reserves for migratory species focus on protecting vulnerable life stages, such as spawning grounds or juvenile habitat, whereas most models simply address lowering fishing mortality (see Chapter 5).

TABLE 6-3 Summary of Studies Estimating Marine Reserve Area Relative to the Conservation or Management Objective

Goal	Citation	Criteria (Species)	Area
Ethics	Ballantine, 1997	Typical terrestrial target	10%
Risk			
Risk management	Lauck et al., 1998	Uncertainty in stock assessment	31-70%
	Roughgarden, 1998	Recruitment overfishing	75%
	Guénette et al., 2000	Spatial model, with and without additional regulations (cod)	20%
	Mangel, 2000	Maintain stock at target levels	20-30%
	Goodyear, 1993	Prevent recruitment overfishing	+20%
	Mace, 1994	Precautionary approach	+40%
	Mace and Sissenwine, 1993	Prevent recruitment overfishing	+35%
	Sumaila, 1998	Bioeconomic model, cost-benefit (cod)	30-50%
	DeMartini, 1993	Yield-per-recruit model, adult mobility (coral reef fish)	20-50%
Risk minimization and bycatch avoidance	Man et al., 1995	Metapopulation model	20-40%
Risk minimization and yield maximization	Soh et al., 1998	Target high biomass areas (rockfish)	4-16%
	Foran and Fujita, 1999	Fecundity and recruitment (Pacific ocean perch)	25%
	Guénette and Pitcher, 1999	Fecundity and recruitment (cod)	+30%

continues

TABLE 6-3 *Continued*

Goal	Citation	Criteria (Species)	Area
Yield Maximization	Pezzey et al., in press	Bioeconomic model (coral reef fish)	21-40%
	Sladek Nowlis and Roberts, 1997, 1999	Fishing intensity (reef fish)	40%
	Sladek Nowlis, 2000	Fishing intensity (Caribbean white grunt)	30%
	Sladek Nowlis and Yoklavich, 1998	Catch enhancement (Pacific rockfish)	20-27%
	Holland and Brazee, 1996	Bioeconomic model (red snapper)	15-29%
	Hannesson, 1999	Bioeconomic model (cod)	50-80%
	Polacheck, 1990	Yield per recruit model/adult dispersal (cod)	10-40%
	Hastings and Botsford, 1999	Reproductive output (sea urchin)	35%
	Botsford et al., 1999	Vulnerability to recruitment overfishing (sea urchin)	8-33%
	Attwood and Bennett, 1995	Increase spawning stocks (recreational surf zone fishing)	33%
	Quinn et al., 1993	Allee effects and dispersal (sea urchin)	50%
	Daan, 1993	Reduce fishing mortality by 10-14% (cod)	25%
Biodiversity			
Representation	Turpie et al., in press	Species representation, complementarity (fish)	10 - 36%
	Bustamante et al., 1999	Representative habitats	36%
	Ward et al., 1999	Habitat and species assemblages	40%
Maintenance of genetic variation	Halfpenny and Roberts, in review	Habitat representation or replication	10%
	Trexler and Travis, 2000	Selective pressure from fishing	20%
Increase Connectivity Among Reserves	Roberts, in review a	Dispersal distance	30%

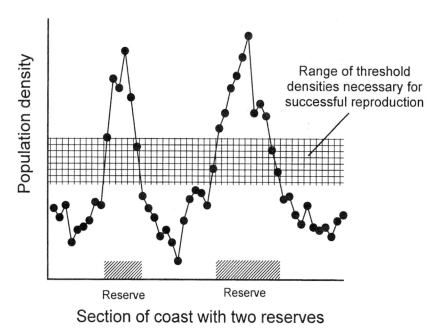

Section of coast with two reserves

FIGURE 6-2 Hypothetical population densities of a marine invertebrate species along an imaginary stretch of coastline with two fully protected reserves. Due to Allee effects at reproduction, the species can reproduce successfully only above a certain threshold of population density, shown as the checked area on the figure. In this circumstance, such densities are reached only inside reserves. Although the species exists outside reserves, only the reserve populations contribute to recruitment.

Although fishery management issues provide much of the justification for reserves, some recent studies have examined more general conservation arguments. Almost all discussions of the design of systems of reserves emphasize the importance of preserving the same species and habitats through replication in several different reserve sites. Several studies that have evaluated how much area needs to be protected to achieve representation and replication conclude that areas in the range of 10% to 35% are appropriate (Table 6-3). Broad species conservation may also require interconnected reserve networks. Interreserve distance provides a measure of connectivity (or the likelihood of interaction among populations in different reserves). Species that depend on other populations for recruitment will require networks of reserves that have high connectivity. Connectivity is also critical for persistence of species that are functionally extinct in areas outside reserves, for instance, when an Allee effect demands high population density for reproductive success (Figure 6-2). To extending this argument; if management outside the reserve is poor, the level of human impact will be greater, and larger protected areas will be needed. Most fishery models and a metapopulation model (Man et al., 1995) show that larger closures are needed to maintain populations if no additional management measures are applied outside the reserves. Beyond predicting that more reserve area will be needed as human impacts increase, we still have no clear guidance as to how the proportion of sea requiring protection will change as the intensity of impact outside reserves increases. It is probable that different habitats will require different levels of protection.

To represent and replicate habitats adequately in a reserve system designed for conserving biodiversity, reserves covering more than 10% of the seas are likely to be required. However, providing insurance against overfishing and protecting nursery areas, spawning aggregation sites, or migration bottlenecks may require reserves covering a larger fraction of a given region.

This review underscores the arbitrary nature of the 20% target figure. Few studies to date have assumed any protection outside reserves. Modeling studies suggest that less area would have to be protected as management outside reserves improves. Even with excellent management of nonreserve areas, a reserve system covering around 10% of the area would improve the conservation of ecological communities, provide insurance against uncertainty, and allow monitoring of natural versus human impacts. With less effective management outside reserves, 20% or more may be needed to achieve conservation goals. In any case, the particular characteristics of the habitat, the exploited species, and the management regime will affect how much area is needed to meet management goals. Optimally, areas targeted for full protection (i.e., ecological reserves) will reside within MPAs to provide a buffer zone that enhances the conservation benefits of the reserve. These MPAs could cover a much larger proportion of the region—for example, the Florida Keys National Marine Sanctuary (FKNMS)—while affording full protection only within specified zones.

How Large Should Individual Reserves Be?

To date, much of the evidence on reserve effects has come from small and isolated reserves that are closed to fishing (consisting primarily of artisanal fishers) in the Philippines (Alcala, 1988; Russ and Alcala, 1996) and even a tiny reserve encompassing only 2.6 hectares of coral reef in St. Lucia (Roberts and Hawkins, 1997). The increases observed in species abundance and biomass indicate, first, that fishing has a measurable impact on marine ecosystems and, second that reserves can be used to facilitate recovery of depleted fish stocks. Although in St. Lucia, even some mobile species benefited from the reserve, species that range widely are likely to gain less from small than from large reserves.

Halpern (in review) analyzed data from studies of 76 reserves that were closed to at least some forms of fishing and looked at the magnitude of effects in relation to reserve size. He found that overall effects (combined across all taxa in each study) on abundance, biomass, body size, or species richness were similar in large and small reserves. Although these findings provide support for using small reserves as a management tool, small reserves will not necessarily meet all management needs. Modeling studies predict that protection of mobile species will increase with increasing reserve size. Also, most of the studies in this review did not evaluate the contribution of the reserves to the populations in the surrounding areas. Since many reserves are established to enhance fishery yields, this is an important criterion for evaluating the effect of size on reserve function.

The size of an individual reserve must be balanced against the mobility of the primary species requiring protection and the need for conservation versus enhancement of the fishery. Relative exchange rates will decrease as the size of the reserve increases (Kramer and Chapman, 1999), and for this reason, conservation goals will be better served by large reserves. However, if fisheries are to benefit from spillover of adults and juveniles, there must be net emigration out of reserves. Therefore, fisheries for species with low to moderate dispersal potential will be better served by smaller reserves spaced out across a management area. To meet multiple conservation objectives, networks must incorporate reserves of a variety of sizes.

To be successful, reserves should be large enough to support the persistence (continued existence) of the species within. Modeling results for reserves of different sizes and species with different dispersal characteristics indicate that persistence is generally ensured if reserve breadth exceeds the dispersal distance of resident species by 1.5 times (Hastings and Botsford, 1999). Hence, a larger reserve is more likely to support the persistence of a greater number of species.

The persistence of long-distance dispersers in reserves may depend on recruitment from elsewhere, either from other reserves or from unprotected areas. The availability of recruits will increase with the size of the regional population. Therefore, recruitment will depend on the regional distribution of suitable habitat, the level of exploitation outside reserves, and the amount of habitat within the reserves (Roberts, in press). Consequently, smaller reserves may be effective when the exploitation of species outside the reserves is well managed or when the proportion of protected area increases (Roberts et al., in review a). Beyond these general predictions, there are no hard-and-fast rules about how large a reserve must be for persistence.

The viability of marine reserves depends on more than biology. It requires adequate enforcement and compliance—greater acceptance of small areas generally translates into higher compliance. If levels of compliance and enforcement decrease as reserves become larger, the effects of increasing size on ecological recovery may be obscured. Hence, the conclusion that size does not influence the magnitude of response (e.g., Halpern, in review) in the protected populations could be misleading.

The relationship between the acceptance and ease of enforcement of reserves will depend on the location. Ballantine (1997) notes that acceptability is lower when current use of potential protected areas is intense. He suggests that reserves should be smaller when they are close to coasts than when they are farther offshore. Furthermore, although small reserves may be easier to enforce where it is easy to identify their boundaries and there are many people to watch them, they will be almost impossible to enforce where boundaries are hard to distinguish and patrols are intermittent. Thus, small reserves will be enforceable near coasts but not farther offshore, whereas large reserves will be less acceptable nearshore but may an be implemented and enforced offshore.

MULTIPLE-USE ZONING OF MARINE PROTECTED AREAS

The primary focus of the first section of this chapter has been fully protected reserves, but the intensity and extent of our activities in the ocean are likely to require broader management approaches that offer different levels of protection. For example, as currently conceived, the National Marine Sanctuary Program has a mandate to ensure harmonious use of resources within its sanctuaries. Under this mandate, it is unlikely to be either feasible or desirable for all of the area within sanctuaries to be fully protected. Nevertheless, larger-scale MPAs, such as the FKNMS, play a critical role in coordinating management. To accommodate the spectrum of different uses in larger MPAs, zoning plans are required. Zoning plans will be needed for all but the smallest MPAs because they avoid unnecessary restrictions and facilitate cooperation between managers and users.

The principal objectives of a zoning plan are usually (Kelleher and Kenchington, 1992)

- to ensure the conservation of the MPA in perpetuity;
- to provide protection for critical or representative habitats, ecosystems, and ecological processes;
- to separate conflicting human activities;
- to protect the natural and/or cultural qualities of the MPA while allowing a spectrum of reasonable human uses;
- to reserve suitable areas for particular human uses, while minimizing the effects of these uses on the MPA; and
- to preserve some areas of the MPA in their natural state undisturbed by humans except for the purposes of scientific research or education.

What Types of Zoning Are Useful in MPAs?

Fully protected reserves within larger MPAs help underpin their biological function and ensure that resources are adequately protected. They are especially useful in giving a high level of protection to core areas, such as sensitive habitats or sites important to vulnerable species. For the reasons given earlier, fully protected ecological reserves should often be permanent features of MPAs. Their conservation and fishery benefits will be greatly diminished if protection is only temporary. Although the placement may have to be adjusted, in concept, ecological reserves should be established with the expectation of permanence. In addition to these fully-protected reserves, zoning plans can be used to separate incompatible activities and provide spatially defined management areas that help protect ecosystem attributes while allowing compatible uses.

The focus of marine management in the United States has historically been on commercial and recreational fishing. The interests of fishers have dominated discussions of how to manage the sea; hence, zoning issues often center on fishery regulations. Certainly, conventional fishery management tools, such as seasonal closures or bans on the use of certain kinds of fishing gear, offer useful tools for zoning. The scope of zoning decisions is becoming broader as other groups have demanded representation of their interest in experiencing pristine marine environments. Among them, the most active are conservation groups, but others are growing in voice and include scuba diving groups, animal rights groups, and scientists. Ecological reserves will secure some objectives of these groups but cannot satisfy all conservation goals. Hence, zones could be created that allow catch-and-release fishing, or that are protected from disturbance by particular types of fishing gear. Zoning to manage compatible and incompatible uses within a large MPA allows for resolution of conflicts between conservation goals and marine resource users, and represents an important tool for meeting the broader goals of coastal zone management (Figure 6-3).

Zoning can be useful as an experimental tool, especially as a component of adaptive management. It can be difficult to determine the relative effects of fishing, environmental degradation, and other human perturbations without large-scale, long-term empirical studies in areas where the suspect activity or most activities have been curtailed. User groups often argue that their activities are not harmful and should not be restricted within MPAs. Recreational users argue that catch-and-release fisheries and diving-related tourism are non-consumptive and should be allowed to continue in a fully protected area. Yet damage to ecosystems may occur from such activities, and opposition may arise if some users believe that the MPA is being designed to reallocate rather than conserve resources. For example, commercial fishers may argue that their access is being restricted to benefit the recreational fishing industry. By utilizing different sets of restrictions for different areas, experimental zoning schemes can help determine the impacts of different activities and avoid potential conflicts over allocation.

Fishers in Australia and off southeast Alaska have argued for a vertical (i.e., depth-specific) zoning scheme in reserves designated to protect features such as seamounts and pinnacles. For instance, an area closed to bottom trawling might

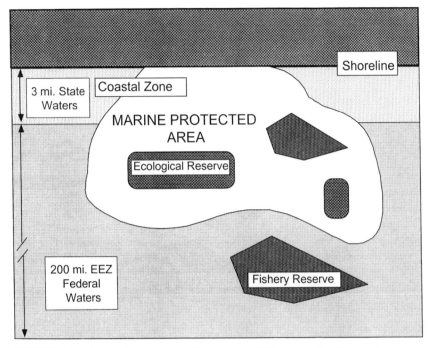

FIGURE 6-3 Schematic diagram indicating how different levels of protection can be applied to zones within a large MPA or used to designate a smaller area (also considered an MPA) to achieve a specific goal. Definitions of MPA, ecological reserve, and fishery reserve are presented in Chapter 1.

NOTE: EEZ = exclusive economic zone.

still allow commercial surface gear such as hook-and-line operations or trolling gear to be used. The Australian seamounts, rising 1,000 to 2,000 m above the seabed, harbor distinct species assemblages found nowhere else in the world. Similarly, two unusually diverse pinnacles off Cape Edgecumbe in southeast Alaska, rising 100 m off the seabed, provide refuge for social aggregations of juvenile rockfish and nesting male lingcod. In each case, the proposed closure prohibits bottom fishing and boat anchoring, thus providing refuges for groundfish and preventing damage to habitat, but allowing surface or mid-water fisheries to continue. Fishers argue that the overlying assemblages are not directly associated with the seamounts or pinnacles and could still support viable recreational and commercial fisheries. Thus, they argue for vertical zoning to allow midslope pelagic fishing. However, it will be important to determine if there are linkages between benthic and pelagic species to ensure that exploitation of surface or midwater fisheries does not harm benthic communities and undermine protection of the reserve.

Part of the problem in vertical zoning is enforcement—it is more difficult to monitor vessels for compliance in a reserve that allows a pelagic fishery than to monitor when fishing vessels are excluded. In the Oculina reserve off Florida, there is evidence that some fishers use modified trolling gear to fish illegally on the bottom (C. Koenig, Florida State University, personal communication, 1999). In such cases, a total closure would be easier to enforce.

Zoning Plans

Zoning plans should consider local use patterns, expectations, attitudes, and knowledge of users with planning undertaken by people closely acquainted with local conditions. Thus, it is not reasonable to expect development of a "one-size-fits-all" model. The format of a zoning plan will vary depending on its legislative basis and local conditions. It may range from a small-scale, locally adopted, municipal plan such as those developed in the Philippines by Alcala and White (1984), to a nationally endorsed legal instrument as required under Australia's Great Barrier Reef Marine Park Act. Whatever the format, most plans will include the following elements:

- statement of the goals and objectives for the planned area as a whole;
- definition of the area with a formal statement of the boundaries of the planned area, a geographic description of its setting and accessibility, and a description of the resources of the area;
- description of activities in the area, concentrating on present uses but in the context of past types and levels of use in the absence of a plan—the description should include social and economic analyses;
- description of the existing legal and management framework applying to coastal fisheries, marine transportation, and other present uses of the area; where they still exist or can be recalled, traditional practices of management, ownership, or rights to the use of marine resources should be described;
- analysis of constraints and opportunities for activities possible within the area;
- statement of the principal threats to the conservation and management of the area;
- statement of policies, plans, actions, interagency agreements, and responsibilities of individual agencies existing or necessary for conservation and management of the area that is to meet the objectives of the MPA and to deal with threats and conflicts—this may usefully include a summary of consultative processes followed in plan development;
- statement of the boundaries, objectives, and conditions of use and entry for the component zones of the planned area;
- provision for regulations required to achieve and implement boundaries and conditions of use and entry; and

- an assessment of the arrangements, including financial, human, and physical resources, required to implement the zones and manage them effectively.

The Planning Program

After an area has been chosen as a potential site for an MPA, there are five desirable stages in the development of a zoning plan (Kelleher, 1999):

1. *Initial information gathering and preparation.* The planning agency, prepares a review of information on the nature and use of the area, including participation by various user groups, and develops materials for public distribution.
2. *Identification of zoning needs.* This involves public comment on the accuracy and adequacy of review materials and discussion of the types of zoning that should be included in the plan.
3. *Preparation of draft plan.* This plan defines specific objectives for each zone and identifies potential sites for these zones.
4. *Publication and review of draft plan.* Public comments are gathered and used to revise the draft plan.
5. *Plan finalization.* The government or agency adopts a revised plan that represents the best fit for meeting conservation needs and concerns of the general public and user communities.

The importance of public participation in planning cannot be overemphasized. Although public participation increases the expense and time involved in planning, it can save both time and expense in the long run by increasing the likelihood that the plan will be approved, implemented, and enforced (Kelleher and Kenchington, 1992). This will require the coordinating agency to identify the various stakeholder groups and develop strategies for working with these groups. For example, planning documents should be accessible, short, and presented in nontechnical, easily understandable language. Participatory mechanisms should not be limited to public hearings but should include less formal settings more conducive to open discussion. Finally, the concerns of stakeholders must be documented in the preparation of management plans (Suman et al., 1999).

In practice, management decisions will be based on incomplete knowledge and understanding; however, most plans involve some research to narrow the range of possibilities for MPA selection. If funds are limited, a competent plan can be developed with basic descriptions of the physical, biological, and socioeconomic characteristics of an area.

A management plan for an MPA will require revisions over time to address shortcomings in the performance of the MPA and advances in understanding of how the MPA contributes to resource management. Hence, some flexibility must be allowed for adjusting the zoning and levels of protection within MPAs. Lack of flexibility currently constrains the National Marine Sanctuary Program because of pledges not to restrict commercial fishing operations made during the designation of some sites (NAPA, 2000). Review of zoning plans and performance should be conducted at intervals short enough for management to respond to problems but not so frequent that it becomes prohibitively expensive—five to seven years is often a suitable period. Such a review should include monitoring impacts, patterns of use, and enforcement. The importance of a rigorous monitoring and evaluation program for the success of MPAs is explored in detail in Chapter 7.

Management Tools to Supplement MPAs

Until recently, the establishment of MPAs has not been viewed in the context of the marine ecosystem as a whole. In planning past MPAs, people have often failed to consider how the surrounding region and human activities may affect the MPA, and vice versa. However, good management of surrounding areas will increase the efficacy of MPAs. With precautionary management, a smaller total area in fully protected reserves may suffice to yield the same conservation benefit. Thus, it is essential to manage activities beyond an MPA's boundaries to secure its long-term viability.

One approach to coupling the use of MPAs with coastal zone management is to create contiguous marine and terrestrial protected areas. Local governments usually have an important role in controlling development and other activities in adjacent coastal areas, as a form of integrated coastal zone management. Because polluted runoff, drainage of wetlands, and diversion of freshwater streams and rivers negatively affect the health of adjacent marine areas, the success of an MPA will in part be dependent on the community's commitment to manage adjacent coastal areas to improve or maintain the quality of the marine environment.

The most obvious external influences that should be controlled include pollution and overutilization of living resources. Effort- and quota-based management, as well as additional spatial and temporal controls on fishing, can be applied to supplement protection using reserves. The type of additional measures needed to regulate fishing will vary widely among fish stocks and areas. Some of these other measures and their application are described in Chapter 3.

Many options will prove useful for spatially controlling fishing outside MPAs, such as establishing area and temporal closures to particular methods of fishing. Temporal closures have long been used in single-species fishery management. In contrast to permanent closures, they have been broadly accepted as a management tool. Temporal closures can be a primary means to try to confine catches to targeted levels or to distribute fishing effort over a season. In fact, within-season temporal closures, combined with gear restrictions, are among the most common methods used to control fishing effort. Temporal closures are also implemented to protect fishery resources at times when they are particularly vulnerable, such as when fish are aggregated on spawning grounds.

MPAs will also shift patterns of fishing effort, and management plans should be designed to take these expected changes into account. Temporal closures aimed at limiting fishing mortality can be effective if regulators correctly anticipate or control the fishery's reaction to time (and area) closures. A closure in one area will typically result in increased effort elsewhere.

In some cases, rotating areas of closure for specified periods has been used to promote growth of young animals, allowing them to reach more valuable sizes. This approach combines temporal and spatial closures to regulate fishing and could supplement benefits from fully protected reserves. As an example, the State of Washington has instituted rotating closures in its sea urchin fishery to control effort and allow populations to recover to marketable sizes and quantities. Rotating closures may help protect habitat and biological communities in addition to target species by allowing areas altered by fishing gear to recover during respites from fishing. However, if the period of closure is short, the area may have insufficient time to recover, jeopardizing the full recovery of the target stock.

Longer-term closures may be instituted as single-species refuges from fishing. In this approach, the goals focus on rebuilding or restoration, with long-term success dependent on more precautionary management after the stock recovers. One recent long-term closure in California, implemented for chinook salmon (*Oncorhynchus tshawytscha*) in the Klamath River area, closed the entire (mixed-species) salmon fishery off the coast to allow recovery. In another example, closure of the severely depleted striped bass (*Morone saxatilis*) fishery along much of the east coast of the United States in the 1980s was effective in promoting recovery of the species and reestablishment of the fishery (Field, 1997; Richards and Rago, 1999). Unfortunately, there are few alternatives to long-term closures of large areas when fish populations have collapsed, and restoration of the abundance and age structure of the depleted population may require many years. One goal of MPAs is to provide insurance against stock collapse (Chapter 5), reducing the need for such drastic measures.

Single-species closed areas lack many of the key conservation benefits of permanent reserves, but they provide an important tool to control fishing effort and supplement more comprehensive MPAs. There are some examples of temporal closures to address multispecies or ecosystem concerns rather than single-species fishery management. For instance, closure of large areas in the Bering Sea around Steller's sea lion (*Eumetopias jubatus*) habitat during the walleye pollock (*Theragra chalcogramma*) fishing season is an example of the use of temporal area closures with broader conservation objectives.

Although temporal closures have value as a tool for fishery management, this approach does not yield the benefits sought with the establishment of MPAs.

There are cogent ecological and socioeconomic arguments for establishing MPAs with long-term, defined boundaries, although there has to be flexibility for adjusting zones, such as ecological or fishery reserves, within MPAs to maximize effectiveness. Damage to habitats and fish populations from human activities can occur very quickly, but recovery often requires a long period of time. For example, biogenic habitat such as deep-sea corals once damaged by trawling could take many decades to recover. Also, when fish stocks collapse, recovery may be slow, especially for long-lived species that take many years to reach reproductive maturity. If reserves are designed in the context of supporting the surrounding ecosystem they can serve as the ecological equivalent of a trust fund: the fish stocks and habitat within a reserve will provide a long-term investment in the productivity of the ecosystem as a whole.

7

Monitoring, Research, and Modeling

Research and monitoring conducted in and around marine reserves and protected areas have three primary and interrelated benefits: (1) better understanding of reserves—how they should be designed and what their benefits and costs are in ecological and socioeconomic terms; (2) deeper knowledge of complex marine ecosystems and the ways that human activities affect these systems; and (3) development and application of marine management methods that are cost-effective in achieving specific goals. Attempts to develop zoning plans for marine protected areas (MPAs) with marine reserves have revealed significant gaps in our understanding of marine ecosystems. Establishment of reserves will provide scientists with opportunities to close these gaps and develop more effective tools for marine conservation.

MONITORING PROGRAMS

Monitoring is an integral component of marine area management; it provides the data required to evaluate changes in marine ecosystems as a result of the implementation of MPAs, especially areas zoned as ecological or fishery reserves. These evaluations are essential for determining effectiveness, improving design, and providing progress reports to stakeholders. Monitoring refers to the periodic evaluation of specific attributes of the ecosystem(s) and socioeconomic conditions represented in or relevant to MPAs. Attributes to be included in a monitoring program will depend on the goals established for the MPA and the main stresses experienced by the ecosystem. Some of the questions that should be addressed through monitoring include the following: Does the MPA system meet its goals and why or why not? Are there unanticipated consequences? Are the size and location of reserves within the MPA optimal? Monitoring programs provide managers with crucial information for evaluating the current status of protected areas and the efficacy of conservation measures. For researchers, monitoring programs provide valuable data that are needed to identify trends in the health of living resources, trends that reveal fundamental features of how ecosystems function and help scientists distinguish between changes that are the result of human influences and those that are natural environmental fluctuations.

Four categories of information may be included in a monitoring program: (1) structure of marine communities (abundance, age structure, species diversity, and spatial distribution); (2) habitat maintenance or recovery; (3) indicators of water quality or environmental degradation (e.g., pollutants, nutrient levels, siltation); and (4) socioeconomic attributes and impacts. For each category it is important that monitoring programs survey sites representative of the MPA, include replicated and comparable sites with different levels of protection, and employ standardized sampling techniques.

General Considerations

Temporal and Spatial Controls for Evaluating Marine Reserves

There are two approaches to analyzing the impacts of marine reserves on living resources. In the first approach, changes within the reserve are evaluated temporally such that conditions are documented before the implementation of protections and then compared to conditions following implementation. A limitation of this approach is that environmental variation in the years before and after the establishment of the reserve may obscure trends resulting from protection. For instance, variable recruitment in a fishery due to a change in oceanic conditions may affect, either positively or negatively, the apparent recovery of a stock after closure of an area. In Kenyan reefs, a twofold increase in fish abundance was observed in surveys of both unprotected and protected sites (McClanahan, 1995); hence, the change was independent of the reserve.

In the second approach, changes in the marine reserve are evaluated spatially such that conditions inside the reserve are compared to conditions in a similar area outside. The limitation of this approach is that reserves often encompass unique habitats; hence, there are few situations in which comparison areas accurately represent the features found within the reserve. For example, in the Polunin and Roberts (1993) study of marine reserves in the Caribbean, differences between fished and unfished areas could have been due to differences in habitat. The site chosen for the reserve might have had higher overall fish abundances even before fishing was halted.

Hence, to test conclusively whether reserves have a particular ecological effect relative to their original goals, it is necessary to establish monitoring regimes at multiple localities that include surveys before and after reserve establishment. Ideally, species survey methods should be rigorous enough to detect a 10-25% change in biomass, density, or species numbers. In many cases, however, such quantitative rigor is difficult to achieve. For example, in surveys of economically important snails on Kenyan reefs, McClanahan (1995) found that the population increases observed in reserves for seven of nine species of snail were statistically nonsignificant because of the overall low density of snails and high variation among sites.

The Need for Systematics

Monitoring species diversity requires knowledge of the systematics of marine species. Taxonomic expertise is necessary to identify the early life stages of various species to assess changes in recruitment success due to reserves. Although the taxonomy of most fish is well known, there are currently few scientists capable of identifying a large number of invertebrate and algal taxa in marine ecosystems. There are well-illustrated atlases for the larval forms of many temperate fish. However, similar resources are not available for the identification of fish larvae in other regions, specifically at low latitudes, or for the identification of fish eggs. Molecular tools to augment and perhaps automate the identification of species are being developed, but even these depend on a firm taxonomic foundation and require input from specialists. A matter of concern in this regard is the decline in the number of trained taxonomists, due to decreased institutional emphasis on systematics, as noted in a prior report (NRC, 1995).

Structure of Marine Communities

Many reserves in the United States and around the world have focused on enhancing or preserving commercially valuable species. Consequently, these populations have been the major or exclusive focus of monitoring (e.g., Davis, 1977; Russ and Alcala, 1989; Smith and Berkes, 1991; Keough et al. 1993; Attwood et al., 1997; Jennings and Polunin, 1997). In reserves designed to enhance biodiversity or stabilize populations of exploited or nonexploited species, monitoring programs have generally emphasized the distribution and abundance of species throughout a wide taxonomic range (e.g., Castilla and Durán, 1985; McClanahan, 1989; Cole et al., 1990; Castilla and Varas, 1998).

Monitoring trophic status (e.g., mean trophic level) of aggregated macrofaunal biomass in the community can provide information on the state of community maturity and potential stability. Such ecosystem-wide monitoring has been instrumental in deciphering the dynamics of marine community interactions, with the result that critical linkages among trophic levels have been revealed.

For example, studies on Kenyan reserves have shown that predatory fish control burrowing sea urchin abundance to such an extent that overfishing has led to an urchin explosion that threatens the physical integrity of reefs (McClanahan and Shafir, 1990). In New Zealand, such trophic interactions impacted even more of the food chain. Because fish are the major predators regulating the population of sea urchins, closing areas to fishing converted them from urchin-dominated algal "barrens" to kelp forest (Babcock et al., 1999). In the Las Cruces Reserve in Chile, protection of a strip of shore from collection of the predatory snail *Concholepas* not only increased the abundance of *Concholepas*, but also caused major changes in the community fauna and flora (Castilla and Durán, 1985). Similarly, strong ecological shifts may be occurring in other marine reserves, especially when the protected species contribute to or support biological habitat complexity associated with kelp, reef corals, or a variety of benthic invertebrates. However, understanding or even recognizing these changes requires broad monitoring of biological communities.

Monitoring programs can detect changes in ecological processes that occur within reserves, in addition to differences in species compositions. Such monitoring may reveal a great deal about community structure and function. Studies in the Leigh Marine Reserve in New Zealand indicated that target fish species can act as keystone predators whose removal causes strong shifts from sea urchin barrens to kelp-dominated ecosystems (Babcock et al., 1999). More recently, the Florida Keys National Marine Sanctuary (FKNMS) has been the site of two studies of ecological processes inside and outside reserves. The first is a set of experiments monitoring the level of herbivory by reef fish where remarkable increases in herbivore pressure have been observed shortly after reserve establishment. The second study monitors recruitment of juvenile corals inside and outside reserves to understand the impact of ecological shifts on the potential of reefs to rebuild (FKNMS, 1999). These ecological studies—focused on important processes, not just species diversity—are crucial parts of a full understanding of reserve function.

In the United States, most monitoring tracks economically important species, especially fish, with the result that ecological information about marine communities is sometimes scarce (Table 7-1). For example, an increase in biomass of fish in reserves may increase predation on invertebrates if they represent the main food source. An increase in predation on small species with quick generation times, such as epibenthic amphipods, may provide an early indicator of ecological effects in newly created reserves. However, such species are rarely monitored, so little is known about changes in abundance of taxa at lower trophic levels.

However, it will be impractical to acquire full species lists and abundances for every marine reserve and protected area, especially because so many marine species are still undescribed (Grassle and Grassle, 1976; Grassle and Maciolek, 1992; Knowlton, 1993). Instead, it will be necessary to emphasize functional

TABLE 7-1   Monitoring Studies in Selected Marine Protected Areas

Region	Taxa Monitored	Reference
South Africa	Fish	Bennett and Attwood, 1991
Fiji	Fish	Jennings and Polunin, 1997
Spanish Mediterranean	Fish	Bayle-Sempere et al., 1994
Northwest Mediterranean	Fish	Bell, 1983
Philippines	Fish	Russ and Alcala, 1989
Northwest Mediterranean	Fish	Garcia-Rubies and Zabala, 1990
Florida Keys	Fish	Ault et al., 1998
New Zealand	Fish, urchins, lobsters	Cole et al., 1990
New Zealand	Fish algae, urchins, lobsters	Babcock et al., 1999
Kenyan Reefs	Fish, urchins, algae, corals, snails	McClanahan and Shafir, 1990
St. Lucia	Urchins	Smith and Berkes, 1991
Dry Tortugas	Lobsters	Davis, 1977
Lac Cruces, Chile	Sessile invertebrates, algae, mobile invertebrates, plankton	Castilla and Durán, 1985
Saba	Fish, corals, gorgonians, sponges, habitat complexity	Roberts, 1996
Belize	Fish and habitat complexity	Polunin and Roberts, 1993
Australian Intertidal Beaches	Snails	Keough et al., 1993
Florida Keys	Corals, algae, lobster, fish, recruitment herbivory	Florida Keys National Marine Sanctuary, 1999
Mafia Island, Tanzania	Corals, fish, algae, environmental quality, habitat, human activities	Agardy, 1997

groups of invertebrates and fish to gain wider coverage of the ecologically important components of ecosystems. For example, fish biodiversity tends to increase with greater habitat complexity. Consequently, monitoring in marine reserves should include documenting changes in structure-forming sessile organisms such as forest-forming algae; reef-forming colonial invertebrates; or bed-forming mussels, oysters, and clams (Roberts, 1995). Since many species vary in their habitat-building potential, such taxa should be cataloged at the species level.

Primary Features of Fishery Monitoring

Fishery reserves are planned and designed to improve management of particular species of fish. The intensity and temporal-spatial scales of the monitoring effort will be dictated by many factors, including the particular objective of the reserve, as well as logistical constraints imposed by reserve size, location, and availability of funds to support monitoring. A well-designed monitoring effort will help to ensure that the status of fisheries and associated resources is known and that the contribution of the fishery reserve is properly assessed (Carr and Raimondi, 1998).

Ideally, fish stocks, habitat, key prey species, and the socioeconomic elements of a fishery will be assessed or monitored prior to reserve implementation to provide a proper baseline for comparison after the reserve is established. Thereafter, regular assessments and determination of stock status for fished species and possibly other key species must be conducted at least annually to determine status and trends within the boundaries of the reserve, but also at its edges and in regions adjacent to it. Many of the fishery-related variables will be the same as those used in conventional stock assessments (Pope, 1988; Shepherd, 1988) and can be used to monitor the status and trends of fish stocks (Fabrizio and Richards, 1996; NRC, 1998a) within reserves.

It is essential to categorize adequately the species complex, abundance, size (used to determine biomass), and age structure of the fished components (fish and invertebrates) as part of the monitoring effort. To provide comparable statistics, fishery-independent data must be collected both within and adjacent to the reserve. Potentially, additional information on location, abundance, and age structure of fished stocks can be obtained from catch-per-unit-effort (CPUE) statistics. In addition to fished species, it is desirable to monitor their key prey and predators with respect to distribution, numbers, and size within the reserve and in adjoining regions. Proper design of the monitoring program is critical to ensure that trends in abundance are detectable. It may be sufficient to monitor relative abundances at age (or size), rather than actual abundances, although each program will have to judge its particular need. Estimating actual abundances and biomasses may be too expensive and, hence, unrealistic for many long-term, fishery-independent monitoring programs.

A determination of spatial and temporal patterns in the distributions of target species is essential. For demersal species, associations with habitat features should be monitored, including shifts in utilization dependent on size and maturation. Propensities for shifts in habitat utilization or spatial distributions (e.g., depth zones occupied) at different life stages are important to evaluate. Monitoring programs that are temporally and spatially intensive may be required to measure emigration from reserves to surrounding regions or to determine if the reserve is acting as a source of recruits for fished areas. An important component of reserve monitoring programs should be the development of recruitment indices, a measure of reproductive success, for target species.

Growth and mortality rates of species that are targeted by fisheries within and outside reserves should be monitored to assess the effects of the reserve on productivity and provide estimates of natural mortality for use in stock assessment models. Egg production, or potential egg production, should be monitored because fecundity of fish populations in reserves should improve if the age and size structure increase.

Fishery Data Derived from Reserves

Established reserves provide a unique opportunity to determine fishery parameters needed to enhance the reliability of fishery management outside reserves. Through targeted monitoring of particular features of populations inside and outside reserves, important information on how fishing impacts the population can be obtained.

Recovery Rates

Immediately upon establishment of a reserve, the trajectory of ecosystem change can provide key information about the rest of the ecosystem that is difficult to obtain in any other way. Larval recruitment into a new reserve indicates the level of potential recovery of species from overexploited population sizes. For instance, failure of reserves along the north coast of Jamaica to rebound in fish population sizes, despite rebounds elsewhere in the Caribbean, suggests a regional collapse of larval availability due to long-term overfishing in Jamaica. Likewise, rapid recovery of certain fish and lobsters in the Florida Keys National Marine Sanctuary shows that aspects of the keys ecosystem can recover from current ecological stresses. Whether corals will return to abundance faster in areas of low human impact is a key question that only the reserves in the FKNMS can provide. It is possible that prior to reserve implementation, environmental degradation had been so severe that recovery is unlikely. Because we do not understand the nature of all of these "points of no return," it is critical to carefully evaluate recovery rates relative to condition of the site at the time it was protected.

Similarly, monitoring of the cod stocks on Georges Bank since the implementation of Closed Areas I and II will provide valuable information on cod population dynamics. Recovery of the stocks may be different within closures that are opened to scallop dredging versus those that remain closed. Understanding this process is crucial to the ability to implement and manage future cod closures or to understand the relationship between cod populations and populations of other important benthic and pelagic species.

Monitoring Export

A key feature of reserves for fishery management or biodiversity enhancement is the ability to export adults, juveniles, larvae, or eggs from within the reserve to the surrounding ecosystem. Understanding the nature and magnitude of this export function will require more extensive research and will require monitoring of reserves and surrounding areas for evidence of such transport.

Bennett and Attwood (1993) showed that fish from reserves moved outside with rates that were specific to different species, but they did not observe net export in these South African systems. McClanahan and Kaunda-Arara (1996) addressed the export issue indirectly by examining fishing records outside a Kenyan reef reserve. They noted an increase in CPUE at the reserve boundary, suggesting a strong export function. However, there was a decrease in CPUE farther away from the reserve and placements of fish traps suggested that local fishers had moved traps to the periphery of the reserve and were apparently catching most of the exported fish. Likewise, Russ and Alcala (1996) saw an overall increase in fish capture in communities surrounding reserves in the Philippines.

Studies of recruitment inside and outside reserves will be needed to derive estimates of export distance that are more direct. Hockey and Branch (1994) showed an exponential decline in recruitment of an intertidal limpet at increasing distances from a reserve in the Canary Islands. Castilla and Varas (1998) showed that spores of the short-distance dispersing kelp *Durvillia* are more common inside the reserve than outside. Stoner and Ray (1996) found that larvae of the queen conch were more common within the boundaries of a reserve in Bermuda, but they did not determine the pattern of drop-off of larval abundances beyond the reserve boundary.

Theoretical predictions of long-distance export on the basis of oceanic current patterns and larval life histories (Roberts, 1997b) suggest that some reserves are likely to be sources for larvae that may recruit far downstream, whereas others may serve as sinks, with no suitable downstream areas for settlement of juveniles. These potential patterns should be tested against observed measurements of average larval dispersal from existing reserves, but such measurements are not yet available. Data on coral reef fish recruitment (Jones et al., 1999; Swearer et al., 1999), and the genetic differences among populations relative to current patterns (Shulman and Bermingham, 1995; Palumbi, 1997; Barber et al., 2000; Cowen et al., 2000) suggest that, in many cases, larval dispersal may be far lower than predicted by oceanographic models of passive particle flow. Hence, direct measurements of larval retention within reserves and the dispersal of larvae outside reserve boundaries will be needed to assess the capacity of reserves to self-replenish and function as sources of new recruits to surrounding areas (Kramer and Chapman, 1999).

Habitat Maintenance and Recovery

In reserves, MPAs, and surrounding open areas, habitat features and status should be monitored to evaluate the potential to sustain diverse and productive communities or fishery landings. For example, the extent and condition of substrates, seagrasses, corals, and live-bottom reef habitats, as well as water quality, should be monitored regularly to determine trends in the condition of MPA habitats. Many other biological and environmental variables can be included in the suite of monitored features or processes (e.g., weather, freshwater inputs, circulation variables, primary production, zooplankton, benthic communities).

Indicators of Environmental Degradation

Many previous reports have described the requirements for effective marine environmental monitoring (NRC, 1990, 2000a, b; Jameson et al., 1998) in more detail than can be offered here. Indicators particularly relevant to MPAs may include dissolved oxygen and nutrients for assessing eutrophication, contaminant loads for specific toxic chemical pollutants, salinity and turbidity for land-based runoff, and chlorophyll for primary productivity. The specific needs will depend on the location of the MPA; for example, in the Florida Keys it would be appropriate to monitor the percentage of live coral cover and water temperature to evaluate the condition of the reefs. In other areas, there may be specific species that could be used as sentinels for a change in environmental quality.

Social and Economic Indicators

Many MPAs are as concerned with tourism as with fishing. Therefore it is important to include monitoring of public perception and use of MPAs to help managers respond to community concerns and adapt regulations to improve their effectiveness. To quote from the Joint Group of Experts on the Scientific Aspects of Marine Pollution (GESAMP, 1996) publication:

> Baselines and monitoring that document public perceptions and governance procedures should be in place from the beginning, so that the social sciences can be applied to overcoming social problems and the effectiveness of governance can be assessed and appropriate actions taken.

For fisheries, the institution of either fishery or ecological reserves will precipitate changes at local and regional scales. It is important to monitor changes in the human dimensions of the fishery so that the cultural and bioeconomic consequences of the reserve can be evaluated. The involvement, effort, and behavior of fishers must be monitored to judge how the reserve has impacted the fishery. The number of fishers, number of vessels, and distribution of the fishing effort in time and space should be monitored at least seasonally. Ideally, this information would be available, or could be obtained, before implementing the reserve to determine how much reserves affect human activities. Monitoring levels and values of catch, amounts of effort directed at key species, and costs of fishing in the vicinity of the reserve will provide the knowledge necessary to determine the social and economic benefits and costs of instituting the reserve. In cases where there are recreational and commercial sectors in a fishery, both must be monitored at local and regional scales, especially if a marine reserve changes the allocation of access to fishery resources among sectors. The economic indicators derived from monitoring will provide information to judge how the value of the fishery has been affected by a reserve or over what time frame the reserve might be beneficial if ecosystem properties were enhanced and fishery productivity improved both locally and regionally. In a cultural and social sense, issues of satisfaction also can be addressed through monitoring efforts, especially through surveys of participants in sectors of the fisheries. The results of monitoring (ecological and socioeconomic) should be made available on a regular basis to ensure that fishers are fully informed about the performance of reserves.

RESEARCH NEEDS

Attempts to design MPAs and reserves have identified major gaps in our understanding of marine ecosystems, gaps that will have to be filled to optimize the use of reserves in marine management. How will new knowledge improve the design and monitoring of MPAs and marine reserves? MPAs and reserves present opportunities for conducting ecological experiments on spatial and temporal scales that have only infrequently characterized studies in marine ecology. Furthermore, basic research in marine ecology and research involving the study of reserves are mutually beneficial. Whereas theories of marine ecology are essential for formulating hypotheses on the optimal design of reserves, reserves provide unparalleled opportunities for testing hypotheses in marine ecology, especially those involving predictions about effects occurring over wide spatial and temporal scales and effects of human impacts on marine ecosystems. Preceding sections of this report have highlighted gaps in our understanding of marine ecosystems. Here, some of these issues are reexamined to emphasize where research efforts are most needed.

Connectivity

Many important shortfalls in our understanding of marine ecosystems, as well as of MPA and reserve design, are related to "connectivity" (see Chapter 6). This concept applies to several phenomena that lie at the very core of marine ecosystem analysis.

Life-History Requirements for Connectivity

What types of habitats do species need during stages of their life histories, and how can the design of MPAs and reserves help ensure that individuals will have access to these habitats? How close together must these habitats be? Research focusing on these questions will contribute to decisions regarding the protection of spawning grounds, benthic nurseries, and pelagic habitats required to support life histories of target species. Because the proximity of different habitats influences the probability of successful transition of organisms through different stages of their life cycles, it is necessary to determine the dispersal potential of each stage (Kramer and Chapman, 1999).

Roles of Oceanography, Physiology, and Behavior in Dispersal

The dispersal ranges of marine organisms may be the most important factor to consider in designing marine reserves (see Chapter 6). Dispersal of young may determine levels of recruitment inside and outside the boundaries of a reserve. To predict regional effects of reserves in particular, we must understand the dispersal of each stage in the life histories of key organisms.

Research is needed on several issues that affect dispersal. The first is measurement of average larval dispersal distances. New approaches to evaluate dispersal include (1) detection of environmental signatures in fish otoliths (Swearer et al., 1999) and crustacean exoskeletons (DiBacco and Levin, 2000); (2) mark-and-release studies using a fluorescent compound to mark larval otoliths (Jones et al., 1999); and (3) determination of patterns of genetic isolation by distance to infer single-generation dispersal distances (Palumbi, 2000). Cowen et al. (2000) noted that simple advection models might overestimate larval dispersal distances, indicating the need to understand larval retention mechanisms and their role in promoting local recruitment.

Critical factors determining larval dispersal potential include (1) current regimes that may determine where a larva or juvenile is transported; (2) behavioral capacities of the organism, which may help to determine whether it is passively carried by currents or able to govern, at least in part, where it settles to become an adult; and (3) physiological abilities, including metabolic rates and level of energy stores, which may determine the potential duration of the larval or juvenile stage (Marsh and Manahan, 1999; Marsh et al., 1999). To understand dispersal, a multidisciplinary research and modeling program is required that blends ecology, genetics, oceanography, behavioral biology, and physiology.

Genetic Connectivity

The high dispersal potential of many marine species has implications for the geographic distribution and genetic connectivity among marine populations. In theory, even low levels of migration among populations will suffice to maintain genetic homogeneity (an indication of connectivity), dependent on the level of selection pressure operating on any given population. Therefore, the genetic connectivity of widely separated populations may be tested by measuring the degree of divergence in their genomes. A high level of divergence indicates that genetic connectivity is low, suggesting limited dispersal. Conversely, low divergence (genetic homogeneity) is less informative because even modest levels of dispersal (in the range of a single migrant per generation) are sufficient to maintain homogeneity. Genome divergence can thus be used to test predictions about the dispersal potential of various species. For example, do species with long-lived pelagic larval or juvenile stages have greater genetic homogeneity across their full biogeographic range than species with brief larval or juvenile stages? Is there enough dispersal to provide sufficient gene flow to prevent the divergence of genomes in widely separated populations? Are species with lower dispersal capabilities characterized by finer-scaled adaptations (i.e., are they relatively specialized for a habitat characteristic at a particular locale—for instance, temperature)? A high degree of specification for a particular habitat may translate into less successful larval or juvenile colonization in widely separated marine reserves.

Shulman and Bermingham (1995) described divergence in mitochondrial DNA of coral reef fish populations in the Caribbean, showing that some populations were significantly differentiated, but others were not. Surprisingly, fish with the lowest dispersal capability did not have the greatest structure, and significant gene flow corridors were not necessarily predictable based on oceanic currents. Likewise, Barber et al. (in press) recently showed strong genetic endemism in Indonesia reef shrimp species, discovering that strong ocean currents do not homogenize genetic structure even over short distances. The history of marine populations since sea-level rise may play an important, though sometimes hidden, role in their ecological exchange from locality to locality.

With the development of molecular techniques for genetic analysis, including very small larval stages, it will be possible to resolve many questions about genetic connectivity among populations. Recent studies indicating that dispersal may be more limited than previously assumed (Palumbi, 2000), confirm that genetic homogeneity is not a reliable indicator of dispersal at the level necessary to sustain reserves in an ecological time frame (Palumbi, in review).

Local Versus Regional Analyses

One reason we know so little about the diverse issues grouped under the heading of "connectivity" is that most research has focused on local effects, not regional influences of reserves. The absence of data documenting regional effects of reserves may, of course, be a sign that these effects are in fact minimal. This could occur where exploitation near a reserve's boundaries eliminates any potentially important regional influences (McClanahan and Kaunda-Arara, 1996). However, the major reason for lack of understanding of regional effects is the absence of research on the influences of the relatively small area within a reserve on the vast area outside. The shortfall in our understanding of regional effects of reserves reflects a general gap in knowledge of how local habitats are coupled to the regional ecosystem. Studies of dispersal of all life stages, including the movement of adults of economically important species in and out of reserves, are required to evaluate the efficacy of MPAs with respect to regional enhancement of stocks.

Studies of Unexploited Species

Because many reserves have been established in response to the collapse of stocks of one or more important exploited species, the focus of research and monitoring within marine reserves has customarily been on a small number of species of economic significance (see Table 7-1). Economic significance and ecological significance do not always coincide. Many organisms necessary for the restoration of natural ecosystem functioning may be ignored when the focus is exclusively on the few fished species. Hence, it is essential that increased attention be given to nonfished species that play important roles in marine ecosystems.

Biodiversity

The effects of reserves on biodiversity at a regional scale are even more poorly understood than the regional effects of reserves designed for fishery enhancement. Even though reserves have usually been created to conserve biodiversity within the protected area, the broader, regional effects of these reserves merit closer study.

Trends in Biodiversity

Biodiversity can be examined at all organizational levels, from the large-scale diversity of ecosystems to the minutiae of genetic diversity within a particular population. In most cases, studies on biodiversity focus on species diversity as the primary indicator of changes at either higher or lower levels of organization. Species diversity may be divided into four components:

1. species richness—the number of species present;
2. species evenness—the relative abundances of different species;
3. species composition—the nature of the species present (i.e., species list); and
4. species interactions—the effects of a species on the composition of the community and its temporal and spatial variation (Chapin et al., 2000).

Reserves established in regions that have experienced minimal exploitation by humans, and thus remain relatively pristine, offer a unique opportunity to study changes in ecosystems and species diversity due to factors other than anthropogenic influences. Long-term research in reserves may allow the effects of natural variation—for instance, in climate—to be clearly delineated from the effects of human activities. An example of the benefits of this type of long-term study of biodiversity is given by research conducted in the biological reserve at Stanford University's Hopkins Marine Station in Pacific Grove, California, which was designated an ecological reserve (except for scientific collecting) in 1931. Surveys conducted at the time the reserve was established provided key baseline data on species composition and abundance. Studies of biodiversity in the mid-1990s documented large-scale changes in species composition and relative abundances that were conjectured to be caused by the warming trend that occurred during this approximately 65-year interval (Barry et al., 1995; Sagarin et al., 1999). Research at the Hopkins site indicates how important a role reserves can play in elucidating natural, that is, nonanthropogenic, changes in biodiversity. Parallel examination of biodiversity in pristine sites and in similar sites that have experienced extractive activities may allow the effects of human impacts on biodiversity to be more clearly delineated.

Studies of Exploited Species

Dispersal of Adults: Tagging Studies

Tagging studies have been used in fishery research to track the movements of fish. These methods can be used to track adults from reserves into outlying unprotected waters to provide a number of types of important fishery data. First, the efficacy of a reserve as a resource for replenishing stocks of commercially exploited species can be established by recording the numbers of marked adults that are captured (or otherwise observed) outside the boundaries of a reserve. Tagging thus allows testing of hypotheses about the "spillover" effects of reserves. The reserve's contribution to stocks in unprotected waters may be evaluated, to a first approximation based on the dispersal distances and number of marked adults recaptured outside the reserve.

Mark-recapture studies may also provide insights into temporal changes within a reserve. If an intensively exploited area is placed under protection, the quality of habitat is likely to vary with time. The fraction of individuals resident within a reserve may change over time, in concert with regeneration of the physical and biological features of the habitat. Properly designed mark-recapture studies conducted within reserves could reveal whether individuals from regions outside the reserve enter and remain.

The use of tagging studies in research and monitoring efforts in reserves will provide more information on dispersal patterns as technologies improve. Although standard tagging techniques will probably continue to be the primary approach, electronic tagging employing microprocessors has increasing potential (Metcalfe and Arnold, 1997; Block et al., 1998b; Cote et al., 1998). The most sophisticated of these electronic tags communicate with satellites, providing data on geoposition in addition to data on vertical movements and thermal history. Because these tags are relatively large, they have been used primarily to track movements of large pelagic fish such as tuna (Block et al., 1998a, b). However, development of smaller versions will encourage wider use. Acoustic telemetry tags also show great promise. They allow relatively precise tracking of fish, as shown by studies of daily and seasonal movements of juvenile Atlantic cod (*Gadus morhua*) (Cote et al., 1998). These tags may not be well suited for tracking dispersal over long periods, however, because the fish must be followed continuously from close proximity.

Chemical marking is another method used to track the movements of fish and identify sites of origin and nursery habitats of a stock. The trace-element composition of teleost otoliths (earstones) contains a distinct "signature" from the chemistry of the local water mass in which a particular layer of the otolith was deposited (Campana, 1999; Swearer et al., 1999; Campana et al., 2000). Larvae developing in coastal waters have different trace-element levels in their otoliths from larvae developing in the open ocean. These chemical signatures will persist throughout the lifetime of a fish, allowing determination of the sites of origin of stocks (Campana et al., 1995) and nursery grounds used during development (Gillanders and Kingsford, 1996). Chemical labeling of otoliths in embryos and newly hatched larvae with tetracycline or other chemical markers yields a distinct, permanent mark that has proven useful to track larvae (Secor et al., 1995; Reinert et al., 1998; Jones et al., 1999) and is applicable to following movements or identifying origins of adults as well.

Density-Dependent Effects on Recruitment and Growth

It is seldom possible to separate effects of fishing on stocks from effects of environmental variability. Networks of fishery reserves could provide a mosaic of fish densities with which to compare how density affects processes such as growth and local recruitment. It is important to determine the life stages most sensitive to density-dependent effects in order to judge whether a reserve might increase recruitment potential locally or through dispersal of prerecruits to adjacent areas that are open to fishing. Fishery scientists generally attempt to link recruitment to adult stock density without consideration of how adults are distributed in space. Research on such processes in reserves will allow efficient, spatially explicit strategies to be developed that will advance understanding of population dynamics over a range of spatial scales.

Models of population growth and regulation frequently include parameters that express impacts of changing densities on egg production, recruitment, individual growth, and survival, but information about effects of density is frequently difficult to obtain. Populations of exploited species inside reserves are in general 30-100% larger than populations outside (Halpern, in review), and such differences in overall density may provide a natural laboratory to evaluate the density dependence of growth and mortality rates. Likewise, the impact of exploited species on other ecosystem components (e.g., prey species and trophic cascades) may be evaluated by comparing across populations at different density. Such studies need to be conducted with proper concern for design and replication, but the mosaic of population densities in a network of reserves and exploited sites may be a powerful tool to elucidate mechanisms of density dependence in marine population dynamics.

Natural Mortality

Natural mortality rate is a key parameter in fish population modeling and stock assessments, but it is notoriously difficult to estimate (Clark, 1999). Specifically, it is hard to separate the total mortality of a fish stock into its fishing and natural mortality components. Reserves could provide areas where fishing mortality rates are essentially zero, greatly facilitating estimation of the natural mortality rate, provided migration is low or can be estimated by tagging studies.

What Types of Research Should Be Allowable in Marine Reserves and Protected Areas?

If we accept the argument presented above that reserves afford unique opportunities for research, it is critical to define the kinds of experimentation that are to be allowed in reserves of different types. A fundamental requirement in all categories of MPAs is that no research should be allowed that might defeat the objectives. It is important, therefore, to tailor the types of research allowed within the MPA to the level of protection intended for each zone. Here, it may be most appropriate to examine allowable research activities in the context of the categories of MPAs delineated by the International Union for the Conservation of Nature and Natural Resources (IUCN, now the World Conservation Union). At one extreme in this categorization are "wilderness" MPAs (IUCN Category I) in which human effects from extractive activities and other forms of perturbation have been minimal. Such wilderness MPAs are inappropriate sites for conducting studies that, either by intent or by accident, could damage the existing natural ecosystem. For instance, evaluation of the effects of trawling or other extractive techniques should not be carried out in MPAs of this class, but completely nondestructive observational studies would be permitted. At the other extreme, MPAs that fit the IUCN's Category IV definition, "Habitat/Species Management Area," may be suitable for resource exploitation. The definition of each category of MPA then carries with it limits to the amount of perturbation allowed.

Ecological research usually involves a certain amount of environmental perturbation. For example, any sample of soft-bottom habitats must be taken with cores that have small impacts, and most studies rely on the retrieval of samples for taxonomic and voucher purposes. Such samples are extremely important because they preserve material for future researchers to evaluate genetic change as well as changes in species composition and abundance. Also, much ecological research depends on experimental manipulations. The same sorts of data are important to fishery scientists who also need to collect samples for life-history, growth, and fecundity measurements. To prohibit such research denies some of the principal value of MPAs. On the other hand, excessive destructive sampling violates the purpose of MPAs.

In any type of MPA in which research is allowed, conflicts may arise between groups of scientists who have different research objectives and experimental approaches. For instance, conflicts seem likely to arise if one research program wishes merely to observe the biota in a reserve, perhaps in order to establish a time series of naturally occurring changes, whereas another group wishes to do manipulative research involving protocols, such as transplantation or clearing of sites to study recruitment. Even if both "observational" and "manipulative" research activities are legally permissible in an MPA, mechanisms for resolving conflicts between researchers with divergent interests will be needed. These considerations argue strongly for the establishment of MPA research oversight committees that would evaluate proposed research with review processes that not only evaluate the merit of the science per se, but also analyze the potential impact of the research on the MPA. Some of the national marine sanctuaries have committees for this purpose (e.g., see information on the Research Activities Panel of the Monterey Bay National Marine Sanctuary at http://www.mbnms.nos.noaa.gov), and the general adoption of such review panels seems warranted in any reserve in which research, beyond monitoring, is to take place.

Research on the Costs and Benefits of Marine Reserves

One of the most critical areas for future research is comparative cost-benefit analysis of conventional fishery management relative to marine reserves alone or with marine reserves as a supplementary tool. Reserves may provide socially important benefits in addition to fisheries that conventional management does not provide. These benefits could include preservation of biodiversity, research and education, nonconsumptive values, recreation, and tourism. However, there have been few studies on either the magnitude of these benefits or the criteria for measuring them. Thus, research on existing and newly created reserves will be required before we will know how reserves compare in a cost-benefit framework against conventional methods. Several types of research will be needed: (1) valuation of nonfishery benefits; (2) development of calibrated simulation models, incorporating spatial information, that forecast biological responses and reflect behavior and values of fishers and other user groups affected by reserves; and (3) empirical studies of reserves used as demonstration projects to involve stakeholders in tests of the value of reserves relative to conventional management approaches.

MODELING

Mathematical Modeling of Reserve Processes

Although there is a well-developed theory of terrestrial reserves (Higgs and Usher, 1980; Gilpin and Soule, 1986; Pressey et al., 1993), a corresponding theory for marine systems has yet to be fully developed (Simberloff, 2000). Marine and terrestrial reserve theories need to be substantially different because the taxa involved have such different life-history traits and because marine systems are usually expected to yield commercial quantities of food from fishing. Terrestrial models often reference island biogeography theory (Simberloff, 1988) and tend to focus on preserving species or habitat richness in reserves, with little or no emphasis on repopulating adjacent areas to support hunting. In contrast, marine reserve models typically focus on single species, with an emphasis on population dynamics under conditions of human exploitation.

Gerber et al. (in review) categorize existing marine reserve models primarily by (1) whether the model is for single or multiple species, (2) what the key life-cycle elements and larval redistribution mode are, (3) what the density-dependent recruitment mode is, and (4) whether adult migration, stochasticity, and rotating spatial harvest are included in the model. They summarize several key results that appear to be general to virtually all types of models. In particular, at a constant level of effort that would otherwise result in overfishing, models indicate that reserves would increase the yield of the fishery relative to conventional management. A primary value of reserves in these circumstances is the higher reproductive capacity of adults protected from fishing. Models suggest that larger reserves would be required for species with high rates of juvenile and adult movement.

A second condition for high efficiency of reserves that has emerged from various models is that the target species should have moderate rates of juvenile and adult movement (DeMartini, 1993). These modeling results appear to apply generally to fishery reserves, but they are based on a very limited set of environmental assumptions. For example, most models assume that all larvae come from a larval pool distributed equally to potential juvenile habitats. In addition, most models do not allow for a number of reserves, and none allow reserves to be of different sizes or embedded in realistic current regimes. Thus, these conclusions from the first simple models are not realistic enough to be used predictively in specific reserve situations.

Currently, the most common form of reserve modeling is to assume that a habitat is divided into a reserve portion where fishing is limited and a portion in which fishing continues. Eggs produced throughout the habitat develop into a common pool of larvae, distributed proportionately into reserve and nonreserve areas. The value of the reserve depends on the redistribution of fishing effort into the nonreserve area and the level of fishing pressure.

Multispecies and Trophic Models

A great majority of marine reserve models focus on the population dynamics of single species. Even though they are complex in their treatment of fishing mortality rates and population subdivision, in most cases, the population size of only one species is considered. Recently, trophic simulations based on the Eco-Path model have begun to incorporate fishery reserves (Walters et al., 1997). Such models try to estimate the impact of reserves on biomass at all trophic levels in an ecosystem, by estimating the flow of biomass from one trophic level to the next with and without fishing pressure. So far, these models do not allow a complex mosaic of fishing efforts and do not include terms to describe the connectivity among different marine populations. They also focus on biomass rather than number of individuals or individual size, so the relationship between these results and typical fishery theories is not yet well known. However, it is clear that single-species models, no matter how complex they are in habitat structure and dispersal capacity, will not capture the community-wide effects of reduced fishing pressure and that efforts to develop multispecies models are one of the next frontiers in modeling the biological properties of marine reserves.

8

Historical Background and Evaluation of Marine Protected Areas in the United States

INTERNATIONAL HISTORY OF MARINE PROTECTED AREAS

The concept of protecting marine areas from fishing and other human activities is not new. In the nonmarket economies of island nations in Oceania (Polynesia, Melanesia, and Micronesia), measures to regulate and manage fisheries have been in use for centuries. These include the closing of fishing or crabbing areas, sometimes for ritual reasons but also for conservation when the ruler decided an area had been overfished or needed protection because it served as a breeding ground for fish that would supply the surrounding reefs (Johannes, 1978). In the broader, global context of conventional fisheries management, Beverton and Holt (1957) provided the first formal description of the use of closed areas in fisheries management. This work was in part inspired by the increase in fish stocks observed in the North Sea after World War II when the fishing grounds were inaccessible because of the presence of mine fields. Since then, fishery managers have used closed areas to allow recovery of overfished stocks, to shelter young fish in nursery grounds, to protect spawning and migrating fish in vulnerable habitats, and to deny access to areas where fish or shellfish are contaminated by pollutants or toxins (Rounsefell, 1975; Iverson, 1996).

During the 1950s and early 1960s, as marine ecosystems became more heavily exploited by fishing and affected by other human activities, the need to devise methods to manage and protect marine environments and resources became more apparent. Over the last 20 years, many ocean areas served as de facto reserves because they were too inaccessible (e.g., too deep, too remote, seabed too rocky), but modern technologies have reduced the amount of unfished area (Bohnsack, 1990; Merret and Haedrich, 1997). To develop a practical response to the need for protecting coastal and marine waters, the international community had to resolve issues of governance of marine areas. Beginning in 1958, the Law of the Sea provided a legal framework to address sovereignty and jurisdictional rights of nations to the seabed beyond the customary 3-mile territorial sea. Four conventions were adopted, the Convention on the Continental Shelf, the Convention on the High Seas, the Convention on the Territorial Sea and the Contiguous Zone, and the Convention on Conservation of the Living Resources of the High Seas.[1] This history is summarized in Table 8-1.

These early conventions were followed by other activities that address marine environmental issues, including the Convention on Wetlands of International Importance Especially as Waterfowl Habitat (1971, known as the Ramsar Convention (http://www.ramsar.org/index.html), and the Convention for the Protection of the World Cultural and Natural Heritage (1972), known as the World Heritage Convention (http://www.unesco.org/who/world_he.htm). In 1972, the Governing Council of the United Nations Environment Programme (UNEP) reviewed the international situation with respect to emerging environmental problems of wide international significance and created the Regional Seas Programme. Action plans were developed with a particular emphasis on protecting living marine resources from pollution and overexploitation through 13 conventions or action plans (http://www.unep.ch), and the first convention entered into force in 1978 for the Mediterranean Sea. In 1983, another regional seas cooperative arrangement, the Caribbean Environment Programme, adopted the Protocol on Specially Protected Areas and Wildlife of the Wider Caribbean Region (SPAW). This protocol calls for a regional network of protected areas in the wider Caribbean to maintain and restore ecosystems and ecological processes essential to their functioning. Specific components of the Caribbean ecosystem are targeted for protection, including coral reefs, mangrove forests, and seagrass beds.

The first conference on marine protected areas was sponsored by the International Union for the Conservation of Nature and Natural Resources (IUCN, now known as the World Conservation Union) in Tokyo in 1975 (IUCN, 1976). The report of that conference called attention to the increasing pressures imposed by man on marine environments and pleaded for the establishment of a well-monitored system of MPAs that were representative of the world's marine ecosystems. Criteria and guidelines for describing and managing marine parks and reserves were outlined and discussed at the Tokyo conference (IUCN, 1976). In 1980, the IUCN, with the World Wildlife Fund and UNEP, published the World Conservation Strategy, which emphasized the importance of marine environments and ecosystems in the overall goal of adopting conservation measures

---

[1] http://fletcher.tufts.edu/multi/marine.html.

TABLE 8-1 A Brief History of Marine Protected Areas (MPAs)

Year or Period	Activity or Event	Significance for MPAs
Historical and prehistory	The closing of fishing or crabbing areas by island communities for conservation for example, because the chief felt the area had been overfished or in order to preserve the area as a breeding ground for fish to supply the surrounding reefs (Johannes, 1978)	Established the concept of protecting areas critical to sustainable harvesting of marine organisms
1950s and 1960s	Decline in catch or effort ratios in various fisheries around the world	At the global level, the need to devise methods to manage and protect marine environments and resources became strongly apparent
1958	Four conventions, known as the Geneva Conventions on the Law of the Sea were adopted. These were the Convention on the Continental Shelf the Convention on the High Seas, the Convention on Fishing, and the Convention on Conservation of the Living Resources of the High Seas	Established an international framework for protection of living marine resources
1962	The First World Conference on National Parks considered the need for protection of coastal and marine areas	Development of the concept of protecting specific areas and habitats
1971	The Convention on Wetlands of International Importance Especially as Waterfowl Habitat (known as the Ramsar Convention) was developed	Provided a specific basis for nations to establish MPAs to protect wetlands
1972	Convention for the Protection of the World Cultural and Natural Heritage (known as the World Heritage Convention) was developed	Provided a regime for protecting marine (and terrestrial) areas of global importance
1972	The Governing Council of the United Nations Environment Programme (UNEP) was given the task of ensuring that emerging	Provided a framework and information base for considering marine environmental issues regionally. MPAs were one means

continues

TABLE 8-1 *Continued*

Year or Period	Activity or Event	Significance for MPAs
	environmental problems of wide international significance received appropriate and adequate consideration by governments. UNEP established the Regional Seas Programme. The first action plan under that program was adopted for the Mediterranean in 1975. The Caribbean Environment Programme action plan was adopted in 1981, and the Cartegena Convention was adopted in 1983, including the Protocol on Specially Protected Areas and Wildlife of the Wider Caribbean Region	of addressing some such issues
1973-1977	Third United Nations Conference of the Law of the Sea	Provided a legal basis upon which measures for the establishment of MPAs and the conservation of marine resources could be developed for areas beyond territorial seas
1975	The International Union for the Conservation of Nature and Natural Resources (IUCN, now the World Conservation Union) conducted a Conference on MPAs in Tokyo	The conference report called for the establishment of a well-monitored system of MPAs representative of the world's marine ecosystems
1982	The IUCN Commission on National Parks and Protected Areas organized a series of workshops on the creation and management of marine and coastal protected areas. These were held as part of the Third World Congress on National Parks in Bali, Indonesia	An important outcome of these workshops was publication by IUCN (1994) of *Marine and Coastal Protected Areas: A Guide for Planners and Managers*
1983	The United Nations Educational, Scientific, and Cultural Organization (UNESCO) organized the First World Biosphere Reserve Congress in Minsk, USSR	At that meeting it was recognized that an integrated, multiple-use MPA can conform to all of the scientific, administrative, and social principles that define a Biosphere Reserve under the UNESCO Man and the Biosphere Programme

1984	IUCN published *Marine and Coastal Protected Areas: A Guide for Planners and Managers*	These guidelines describe approaches for establishing and planning protected areas
1986-1990	IUCN's Commission on National Parks and Protected Areas (now World Commission on Protected Areas) created the position of vice chair, (marine), with the function of accelerating the establishment and effective management of a global system of MPAs	The world's seas were divided into 18 regions based mainly on biogeographic criteria, and by 1990, working groups were established in each region
1987-1988	The Fourth World Wilderness Congress passed a resolution that established a policy framework for marine conservation. A similar resolution was passed by the Seventeenth General Assembly of IUCN	These resolutions adopted a statement of a primary goal, defined "marine protected area," identified a series of specific objectives to be met in attaining the primary goal, and summarized the conditions necessary for that attainment
1994	The United Nations Convention on the Law of the Sea (UNCLOS) and the Convention on Biological Diversity (CBD) came into force. UNCLOS defines the duties and rights of nations in relation to establishing exclusive economic zones measuring 200 nautical miles from baselines near their coasts. While facilitating the establishment and management of MPAs outside a country's territorial waters, UNCLOS does not allow interference with freedom of navigation of vessels from other countries	These two international conventions greatly increase both the obligations of nations to create MPAs in the cause of conservation of biological diversity and productivity and their rights to do so. It is notable that the United States has not ratified either convention. The Conference of Parties of the CBD has identified MPAs as an important mechanism for attaining the UNCLOS objectives and intends to address this matter explicitly in the next few years
1995	The Great Barrier Reef Marine Park Authority, the World Bank, and the IUCN published *A Global Representative System of Marine Protected Areas* (Kelleher et al., 1995)	This publication divided the world's 18 marine coastal regions into biogeographic zones, listed existing MPAs, and identified priorities for new ones in each region and coastal country
1999	IUCN published *Guidelines for Marine Protected Areas*	These updated guidelines describe the approaches that have been successful globally in establishing and managing MPAs

SOURCE: Modified from Kelleher and Kenchington, 1992.

to ensure sustainable development (IUCN, 1980). In 1982, the IUCN Commission on National Parks and Protected Areas (CNPPA, now the World Commission on Protected Areas [WCPA]) organized a series of workshops at the Third World Congress on National Parks in Indonesia to promote the creation and management of marine and coastal protected areas. This workshop resulted in the publication of *Marine and Coastal Protected Areas: A Guide for Planners and Managers* (Salm and Clark, 1984).

Other activities that recommended implementation of marine protected areas (MPAs) for marine conservation included the First World Biosphere Reserve Congress (UNESCO, 1984), the Commission on National Parks and Protected Areas (IUCN, 1987), and the report *Our Common Future* published by the World Commission on Environment and Development (WCED, 1987). In November 1987, the General Assembly of the United Nations welcomed this report and adopted the *Environmental Perspective to the Year 2000 and Beyond* (UNEP, 1988).

In response to the growing awareness of problems in marine ecosystems, the World Wilderness Congress and the IUCN passed resolutions that established a policy framework for marine conservation. These resolutions stated the primary goal of marine conservation, defined "marine protected area," identified a series of specific objectives to be met in attaining the primary goal, and summarized the conditions necessary for that attainment. They formed the framework for the IUCN policy statement on MPAs that appears in *Guidelines for Establishing Marine Protected Areas* (Kelleher and Kenchington, 1992).

Two international conventions came into force in 1994, which greatly increased both the obligations of nations to create MPAs in the cause of conservation and their rights to do so. They are the United Nations Convention on the Law of the Sea (UNCLOS) and the Convention on Biological Diversity (CBD) (UN, 1983; UNEP, 1992). Proceedings on UNCLOS commenced in 1972, and it entered into force in 1994. Although the United States has not ratified either convention, there is considerable agreement, at least in principle, with many recommendations of the two conventions, and the components of treaties in force often become customary international law. UNCLOS establishes duties and rights of nations to establish exclusive economic zones (EEZs) extending 200 nautical miles from baselines near their coasts. While facilitating the establishment and management of MPAs outside a country's 3-mile territorial waters, UNCLOS forbids interference with freedom of navigation by vessels from other countries. CBD increases the obligations of signatory nations to protect biodiversity, including biological productivity. MPAs have been identified as an important mechanism to attain the objectives of the CBD.

There has been considerable progress in establishing MPAs over the past three decades. In 1970, there were 118 MPAs in 27 nations. By 1994, the number had expanded tenfold to at least 1,306 MPAs in many nations, with numerous other proposals under consideration (Kelleher et al., 1995). It is clear that the foundations for broader implementation and use of MPAs to conserve biodiversity and to promote ecologically sustainable development have been established, both internationally and in many individual countries.

MARINE PROTECTED AREAS IN THE UNITED STATES

Over the last century in the United States, federal, state, and local governments have established parks, reserves, wildlife refuges, and other areas adjacent to and sometimes including marine waters. In addition, many military reservations restricted for other purposes have served over time as de facto MPAs. Currently, the United States is involved in discussions on possible multinational MPAs with Canada, Mexico, and Russia.

In the 1920s, increased interest in marine sciences led to the establishment of small scientific research reserves. For example, the Friday Harbor Laboratory (Washington State) designated the Marine Biological Preserve in 1923 but had limited authority for enforcement. Since then, the National Park Service has established several parks with marine components, including Everglades National Park (1934), Fort Jefferson National Monument in the Dry Tortugas (1935), and Key Largo Coral Reef Preserve (1960). At the state level, California established the Point Lobos Marine Preserve (1960), Florida dedicated John Pennecamp Coral Reef State Park (1960), Hawaii established the Hanauma Bay-Kealakekua Bay Marine Life Conservation Districts (1967), Massachusetts created the Cape Cod Ocean Sanctuary (1970), and Washington State extended the boundaries of nine state parks to encompass adjacent marine areas. Local city governments established underwater parks, such as the 1970 designations of the La Jolla Underwater Park (San Diego, California) and the Edmonds Underwater Park (Washington State).

The foregoing history captures the character but not the detail of MPA designations in the United States up until 1970. Since then, government programs such as the National Oceanic and Atmospheric Administration's (NOAA's) National Marine Sanctuaries Program have extended the number of MPAs. In addition, areas that have fishing restrictions (zoning) with respect to gear, season, bycatch reduction, and other factors also may be considered types of marine protected areas, encompassing considerable portions of the U.S. EEZ.

Relatively few MPAs in place in the United States qualify as marine reserves (NRC, 1999a). National parks presently engaged in reducing or eliminating fisheries (e.g., Glacier Bay National Monument, Channel Islands National Park) are facing significant opposition because fishing traditionally has been permitted within their boundaries. Few of the national marine sanctuaries have policies to curtail or control fishing, although more areas may be designated with restrictions on fishing activities as the sanctuaries develop new general management plans (Davis, 1998).

No overarching legislation, such as the Coastal Zone Management Act

(CZMA) of 1972 (P.L. 92-583), exists for MPAs in the United States. A comprehensive review of each of the many types of marine reserves and the legal and institutional frameworks for their management has not been performed nationally, although efforts to inventory MPAs in some areas have been completed (McArdle, 1997 [California]; Murray and Ferguson, 1998; Mills, 1999; Robinson, 1999; [Washington State]) or are under way (Marine Fish Conservation Network and Center for Marine Conservation, 1999). The following sections present a brief review of the major types of MPAs in the United States, with respect to legislated purpose, management, and attainment of goals. Also included are laws that promote protection of marine habitats or that prescribe management conditions for areas surrounding MPAs (Table 8-2).

Criteria for Evaluation

Analysis of the current system of MPAs is constrained to a qualitative rather than a quantitative evaluation by the limited availability of data. The following sections evaluate MPAs designated by major federal programs with respect to their intended purposes and authorities. The status of MPAs in the United States is assessed with respect to the four broad goals identified for marine protected areas (Chapter 2): conserving biodiversity and habitat, managing fisheries, providing ecosystem services, and protecting cultural heritage. In addition, there is the national goal of establishing an interconnected network of MPAs that represent the variety of marine ecosystems in the United States. In all cases, MPAs have to be adequately monitored, allow for research, be enforceable, and have significant stakeholder involvement.

Although the United States is comparatively well represented in global surveys of existing and potential areas (Kelleher et al., 1995), there is no comprehensive inventory of MPAs in the United States, and there is little information on or analysis of the goals, authorizing legislation, agency role, designation process, and current regulations. The lack of a systematic inventory is perhaps a telling commentary on the fragmented approach to establishing MPAs in the United States. Existing MPAs have been instituted for many reasons using diverse authorities with varying degrees of administrative support and limited funding for monitoring and enforcement.

At the federal level, NOAA has begun an effort to inventory coastal and marine protected areas in the United States. However, to date, this inventory includes land areas that border on but do not contain marine waters and does not include areas designated under fishery management regulations (see Table 8-2). In addition, there are descriptive or educational overviews of protected areas, for example, national marine sanctuaries (Seaborn, 1996; Earle and Henry, 1999) and national seashores (Wolverton and Wolverton, 1994). State-level inventories are available for a few states—California (McArdle, 1997), Oregon (OOPAC, 1994), and Washington (Murray and Ferguson, 1998; Mills, 1999; Robinson, 1999). Al-

TABLE 8-2 Preliminary Inventory of Coastal and Marine Protected Areas in the United States[a]

	Number of Sites	Area (acres)
U.S. MPAs		
Federal	390	149,742,686
State	736	2,535,715
Nongovernmental organizations	128	213,275
Federal Sites		
NOAA	33	11,923,332
National Marine Sanctuaries[b]	12	11,502,720
National Forest Service	114	38,073,257
U.S. Fish and Wildlife Service	42	60,211,701
National Park Service	201	39,534,396
Total	**390**	**149,742,686**

NOTE: The U.S. federal government manages about 700 million acres of land and sea for the purpose of natural resource protection and use; approximately 20% is in coastal and marine areas, and less than 2% is strictly marine.

[a] Data as of 1998. This inventory does not include areas designated under fishery management regulations by federal or state fisheries authorities for fisheries management. Inclusion of ocean and coastal areas subject to fisheries regulations as MPAs in this inventory would change the results significantly. These data are being gathered and will be added to the inventory at a later date.

[b] National Marine Sanctuaries fall under the jurisdiction of NOAA. Therefore, these numbers are represented in the overall NOAA totals.

SOURCE: Lani Watson and Roger Griffis, National Oceanic and Atmospheric Administration, personal communication, 2000.

though the state-level reviews are largely descriptive and were not intended to evaluate performance, they provide a foundation for more detailed examinations.

The most imposing barrier to a systematic evaluation of MPA performance in the United States is the shortage of baseline monitoring of physical and biological parameters within MPAs before and after designation. In many cases, the MPAs are too recently established for significant change to be detected. Even where change is observed, it is difficult to discern cause and effect because management actions, environmental variability, and other exogenous and endogenous factors affect outcome (Ticco, 1996; Allison et al., 1998; Rose, 2000). Thus, performance is difficult to evaluate based on output parameters (Williams, 1980) such as statistically significant increases in fish abundance, stock structure, or species composition of assemblages. Instead, input parameters such as level of funding or program activities are often used as metrics of performance, although they do not measure progress toward management goals.

Description and Evaluation of Existing MPA Programs

National Parks

Areas within the national park system are designated by Congress for preserving unique or pristine scenic and wildlife resources in the United States. The National Park Service (NPS) administers national (marine) parks, national recreation areas, national monuments, national seashores, and national sand dunes. Managing NPS resources often involves a cooperative relationship with state agencies across park boundaries (Fagergren, 1998) to ensure effective resource protection.

There are 201 national parks administered in coastal areas, 30-40 of which have significant marine areas as a component. These areas are managed under provisions of the National Parks Organic Act 1916 (as amended) (16 U.S.C. 1) to preserve natural features unimpaired for future generations, while providing for public enjoyment (Keiter, 1988). Hence, they do not necessarily provide representation of different ecosystems. NPS has the regulatory authority to provide a high level of protection to conserve park resources, consistent with the establishment of marine reserves. For instance, federal regulations prohibit the take of most wildlife, prohibit commercial fishing except as permitted by statute, and unless otherwise specified, follow state-level recreational fishing regulations. In most coral reef areas, such as Biscayne National Park, NPS has a legislative mandate to allow recreational and commercial fishing and shellfishing. In American Samoa (the National Park of American Samoa), however, only subsistence fishing is permitted and no public recreational fishing is allowed (NPS, 1998).

NPS is under pressure to reduce or eliminate commercial and recreational fishing inside NPS-administered boundaries (Davis, 1998), including Glacier Bay National Park, Everglades National Park, and Channel Islands National Park (Kronman, 1999), particularly in cases where fishing impairs the very resources that NPS is charged to protect. The authority of NPS to establish more restrictive fishing regulations has been upheld in Everglades National Park (*Organized Fishermen v. Watt*) (Mantell and Metzgar, 1990), but it has been difficult to garner political support for altering established use patterns. In some parks, the designating legislation specifically exempts activities such as fishing, which are increasingly seen as incompatible uses from the standpoint of nature preservation. Also, because NPS often lacks the baseline data needed to demonstrate resource degradation, fishing interests have successfully argued against restrictions on current fishing practices. Preservationists have argued that the NPS mandate to maintain ecosystem integrity is inconsistent with extractive uses of any kind, regardless of whether or not harm can be proved (McClanahan, 1999).

National parks with a marine component support some of the ecosystem services that appertain to their role as MPAs. Also, to the extent authorized, national parks protect the cultural heritage contained therein. Few of the national parks are located to provide an ecologically interconnected set of reserves, but in several cases (Ferguson, 1997), parks cooperate with neighboring national marine sanctuaries to increase their effectiveness. Research in the marine components of the national parks is supported under a variety of NPS-funded and non-park-funded programs but not at the scale required to meet needs. Although NPS maintains its own enforcement capabilities, the parks rely on education rather than fines or prosecution to obtain compliance. Long-term ecological monitoring is almost nonexistent at the relevant social, economic, and ecological scales. Only recently, has NPS started to develop plans to implement large-scale ecosystem monitoring for its natural resource-oriented units. However, there is insufficient funding to accomplish the task. [2]

The designation process for national parks tends to be a highly political, combining top-down and bottom-up processes. The agency, prompted by local and national interests, performs studies that form the basis for legislative proposals that usually receive public hearings and other forms of involvement. Action to increase the restrictions on historical uses of park areas have been contentious, not only with regard to fishing closures, but also over what constitutes adequate compensation to established fisheries as seen in Glacier Bay (Box 8-1). In contrast, under the lead of the Florida Keys National Marine Sanctuary, NPS and other federal and state entities are using an extensive stakeholder planning process for designing and implementing the Dry Tortugas Ecological Reserve, an approach that has defused some of the controversy (see following section). In the National Park of American Samoa, NPS is endeavoring to preserve a pristine tropical ecosystem consistent with Samoan culture (Chadwick, 2000).

National Marine Sanctuaries

Title III of the Marine Protection, Research, and Sanctuaries Act 1972 (16 U.S.C. 1431-1434) allows the Secretary of Commerce, after consultation with other federal agencies and responsible state officials, to designate national marine sanctuaries (NMSs) for the purpose of "preserving or restoring such areas for their conservation, recreational, ecological, or esthetic values." Sanctuaries also may be designated by Congress. National marine sanctuaries are administered by NOAA's National Ocean Service. The sparse language of the act has been amplified in the development of regulations to include areas of human use value, coordination of management to complement existing regulatory authorities, support of scientific research, enhancement of public awareness and understanding, and facilitation—to the extent practicable—of all public and private uses of the resources not otherwise prohibited (Thorne-Miller and Catena, 1991).

---

[2] www.nps.gov/glba/learn/preserve/projects/index.htm.

BOX 8-1
Glacier Bay National Park and Preserve: Phasing Out Fishing

Glacier Bay,[a] a 3.3-million-acre unit of NPS, is representative of the significance and difficulties of restricting fishing under NPS mandate to preserve natural processes and ecosystems (GBNP, 1998). Glacier Bay's 600,000 acres of marine waters make it the largest marine area under NPS management. Some 53,000 acres of water are designated as wilderness. Commercial fisheries continue in parts of Glacier Bay, even though this has been prohibited since 1966 within the National Park and since 1980 in designated wilderness. Fisheries have been specifically allowed in the preserve since 1980.

NPS objectives for Glacier Bay to enhance park resources and values are

- to preserve and perpetuate habitats, natural population structure, and species distribution;
- to ensure that natural, successional, and evolutionary processes occur unimpeded;
- to ensure that natural, biological, and genetic diversity is maintained;
- to minimize visitor and vessel use conflicts;
- to protect wilderness values;
- to sustain and strengthen Hoonah Tlingit cultural ties to the park; and
- to expand our knowledge and understanding of marine ecosystems.

Commercial fisheries (salmon trolling, halibut longlining, salmon seining, and crab pot fishing) existed in the area encompassed by the Glacier Bay National Monument established in 1925. Specific regulations were developed for these activities in 1939 by the Bureau of Fisheries and in 1941 and 1959 by the U.S. Fish and Wildlife Service. Until recently, the general approach has been to allow established uses to continue.

NPS's failure to close commercial fishing was challenged several times by user groups beginning in 1990. After publishing a proposed rule to eliminate commer-

---

For the most part, offshore oil exploration and production have been the only prohibited activities in the regulations governing most of the national marine sanctuaries, although in some areas such use continues inside and adjacent to the sanctuary.

A total of 13 national marine sanctuaries had been designated under this program as of November 2000 (see Figure 8-1). They range in size from Fagatele Bay NMS (American Samoa) (0.25 nmi^2) and Monitor NMS (North Carolina) (1 nmi^2) to the extensive Olympic Coast (Washington State) (3,310 nmi^2) and Monterey Bay sanctuaries (California) (5,328 nmi^2) (NOAA, 1997). Despite the name "sanctuary," they are better characterized as multiple-use resource management areas (Clark, 1998). The average person is surprised to learn that a wide range of consumptive and nonconsumptive uses occurs within

cial fishing in the park in 1991, the State of Alaska requested that NPS and the Alaskan congressional delegation consider a legislative approach to resolving the issue. No agreement was reached, however. Therefore, NPS proceeded to develop proposed regulations on commercial fishing over a 15-year period within nonwilderness areas of the park. NPS proposal invigorated congressional response and brought legislated modifications in 1998 through the appropriations process (36 CFR part 13). These actions limited the range of NPS actions somewhat but were generally supportive of a phase-out of commercial fishing over the lifetimes of the current participants. With respect to Dungeness crab fisheries, a compensation program was designed to ease the transition for fishermen. It provided an average of $400,000 to each crabber to give up his or her fishery—a total of $8 million. In addition, halibut, Tanner crab, and salmon fishermen with lifetime tenure under the 1998 legislation were allocated $23 million to compensate for not being able to sell their quota shares or limited entry permits (Baker, 2000).

Despite more than 10 years of active discussions and policymaking, the disputes are still unresolved. In August 2000, the U.S. Supreme Court agreed to adjudicate whether the federal government or the State of Alaska should control the waters of Glacier Bay. In brief, NPS argues that it controls the waters because it was the area manager before Alaska achieved statehood in 1959 and that the underwater territory was not transferred to the state. The state regards the waters as under its jurisdiction.

The details of this controversy are fascinating as well as frustrating to observers of efforts to make protection mandates work through the designation of MPAs. They illustrate the complexity of coordinating management objectives, authorities, and realities with respect to competing demands from user groups and changing public values with respect to commercial fisheries in NPS areas. In other areas, the same questions arise with respect to recreational fisheries.

---

[a] The Glacier Bay is also designated as an International Biosphere Reserve and a World Heritage Site.

SOURCE: GBNP, 1998.

national marine sanctuaries. Frequently, protective measures and management within sanctuary boundaries depend on a cooperative or partnership relationship with resource managers from other jurisdictions (Ferguson, 1996, 1997). A panel of the National Academy of Public Administration (NAPA, 2000) recently undertook a review of the National Marine Sanctuaries Program. The panel concluded that sanctuaries should take more steps to protect marine resources within their boundaries, including regulating and prohibiting fishing or other activities when appropriate.

The NMS programs contain a variety of marine environments that constitute neither a representative system of MPAs nor a network. In relation to the objectives and regulatory authority granted to the program, implementation has been successful, with considerable public support for exclusion of oil and gas devel-

158

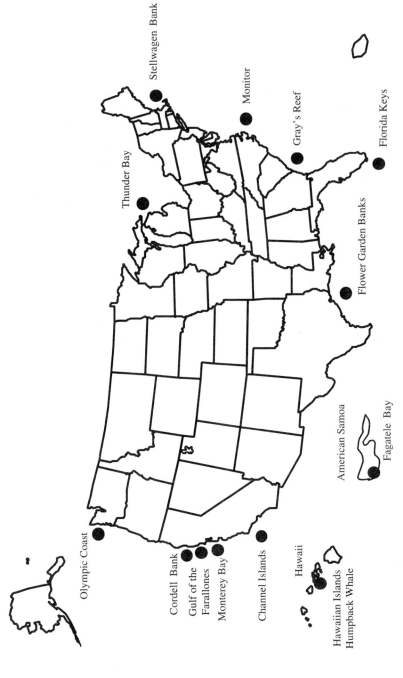

FIGURE 8-1 Map of the 13 national marine sanctuaries.

opment as well as dredging, placement of structures, and dumping (NAPA, 2000). In terms of broader protection mandates, the sanctuary programs facilitate conflict resolution among users (NRC, 1997; Suman, 1997; Suman et al., 1999).

The concept of marine zoning is gaining interest in the form of integrated coastal management (Cicin-Sain and Knecht, 1998; Klee, 1999) in which sanctuaries are a component of the zoning plans at state and national levels. Current development and revisions of sanctuary general management plans incorporate the concept of zoning (Clark, 1998; Salm and Clark, 2000). The Florida Keys National Marine Sanctuary (FKNMS) is a good example of the use of zoning within such a plan (NOAA SRD, 1996b). In this case, Congress required that zoning be used in the FKNMS to develop management plans for the area, in cooperation with other federal agencies and state and private interests in Florida (Suman, 1997). The Monterey Bay National Marine Sanctuary used zoning techniques to reduce conflicts among personal watercraft, swimmers, and beach goers. The Channel Islands and Olympic National Marine Sanctuaries employ special area management for shipping, using vessel exclusion zones and specified vessel traffic schemes to reduce the risk of spills and accidents from vessel transportation (NRC, 1997). Currently, zoning is being considered for designating fishing areas and reserves.

With respect to fisheries, many are looking to the NMS program to increase restrictions on fisheries and to implement reserves as part of its mission. This places the sanctuary programs in an awkward position because at the time some sanctuaries were designated, agreements were struck stating that restrictions on fishing in the sanctuaries would not be imposed by federal agencies. Many of the original sanctuary management plans contain this commitment (NAPA, 2000). Sanctuary managers must work with regional fishery management councils and state fisheries officials to respond to specific measures required for managing fisheries. As an example, Channel Islands National Park and Channel Islands National Marine Sanctuary are currently engaged in general management planning processes that could lead to various restrictions, including possible "no-take" zones, on recreational and commercial fishing for rockfish and other species.

FKNMS, in a joint effort with the State of Florida, the Gulf of Mexico Fishery Management Council, and the National Marine Fisheries Service (NMFS), is proposing a no-take ecological reserve to protect the remote coral reef area of the Dry Tortugas. Studies on the biology and oceanography of the area suggest that this site could serve as a source of larvae of fish, lobsters, and other species for the Florida Keys and the east coast of Florida. Controversy over the original proposal at the time the FKNMS was designated delayed implementation, but recent developments have cleared the way for a collaborative action that satisfies the concerns for coral reef preservation while balancing the economic concerns of commercial fishermen.

Agreement on how to manage the Tortugas was hammered out through a process convened by FKNMS that brought together 25 commercial and recreational fishers, divers, conservationists, scientists, citizens, and representatives of government agencies into a working group charged with using the best available scientific information to evaluate alternative approaches. The working group came up with an ecosystem approach to resolving the problems by focusing on natural resources and not on jurisdictions. Despite previous animosities, this group was able to come to unanimous agreement on a proposal to expand the boundary of the FKNMS by 96 nmi^2 and to establish a two-section Tortugas Ecological Reserve totaling 151 nmi^2.

This agreement has been incorporated into the FKNMS proposal and is currently awaiting approval from NPS and the State of Florida. The fact that the process was inclusive and focused on the dual interests of providing for coral reef protection and at the same time not unduly affecting other user groups demonstrates that collaborative, consensus-oriented processes provide effective mechanisms for developing viable management options in areas with high levels of conflict (see Chapter 4, Box 4-2).

The NMS program has been successful in increasing the public profile of the sanctuaries and increasing public awareness of the nation's marine resources and conservation needs through effective public outreach and education programs, such as the Sustainable Seas Expeditions. One indication of increasing interest in the NMS program can be inferred from access statistics for the NMS Web site. The number of requests (or hits) per month over a three-month period (May, June, July) was 225,520 in 1999 compared to 573,520 in 2000, representing a 2.5-fold increase. Changing values and rising public expectations concerning the role of "marine sanctuaries" as true protected areas are bringing demands to increase the level of protections in the sanctuaries. This interest has been reflected in an elevation of the program's status within NOAA and a 60% increase in funding for FY 2000. Several sanctuaries are in the process of revising management plans, and research and monitoring programs are being proposed with the hope that funding will be available. In 1999, the NMS budget allocation amounted to about $800 per square mile under its jurisdiction. This compares to $6,167 per square mile of U.S. Forest Service jurisdiction and $16,667 per square mile under NPS jurisdiction. Significantly increased allocations in FY 2000 (about $1,250 per square mile) will help NMS managers improve the coverage of monitoring programs and decrease reliance on volunteer efforts. Because of the lack of effective monitoring in most NMS areas, most measures of sanctuary program success are limited to inputs (budgets and activities) (NOAA SRD, 1996a; NOAA, 1997; NAPA, 2000) rather than outputs (resource evaluations).

National Estuarine Research Reserves

The National Estuarine Research Reserve (NERR) is a hybrid program administered at the federal level to provide the funding, infrastructure, and coordination for individual reserve sites that are designated and managed at the state level. By 1999, a system of 25 NERRs had been established under Section 315 of the CZMA. This program is administered by the U.S. Department of Commerce, under NOAA's National Ocean Service, with management authority delegated to state and local governments. Because these valuable estuarine habitats fall under the jurisdiction of the states, they require protection by state law (http://www.ocrm.nos.noaa.gov/nerr/welcome.html).

The primary purpose of the NERR program is to promote and coordinate scientific research, but commercial development of the area is either prohibited or controlled. Although the intent is to protect estuaries for long-term research and education, goals of estuary restoration and recovery also are supported (Thorne-Miller and Catena, 1991). Many different habitat types are included among the NERRs designated to date, but selection has been based more on opportunity than on representation. Monitoring is conducted on a more systematic basis than for most other protected areas but often as a result of scientific monitoring for research purposes or through volunteer activities (T. Stevens, Manager, Padilla Bay National Estuarine Research Reserve, personal communication, 1999). Monitoring has been relatively comprehensive and consistent in the areas of water quality and atmospheric conditions. The NERR System-Wide Monitoring Program (SWMP) program has developed monitoring protocols, trained personnel, and purchased and deployed equipment to increase understanding of how environmental factors influence estuarine change and functioning (Wenner and Geist, 2001).

National Wildlife Refuges

Hundreds of coastal and marine national wildlife refuges (NWRs) dot the shores of the United States. NWRs with marine components contain the largest area under current federal designations. The system is administered by the U.S. Fish and Wildlife Service (FWS) under a variety of laws. Since the early 1890s, presidential declarations and later congressional mandates have been used to designate NWRs that encompass many kinds of wildlife and habitats in the marine environment. Designation relies heavily on the allocation of parts of federal lands or lands acquired by donation or purchase as habitat for wildlife.

Although they are called refuges, hunting and commercial fishing are approved activities in NWRs, but the FWS can institute suitably restrictive measures if required to conserve wildlife or habitat (Adams, 1993). Refuges play an effective role in protecting threatened and endangered species and preserving habitat for wildlife. In the marine realm, this approach has been applied for migratory birds. Concerted efforts have been made to provide for the needs of waterfowl by obtaining representative habitats for resting and feeding spaced appropriately along the migratory flyways. In addition, the nesting and rearing habitats at the terminus of their migrations are also targeted. In this sense,

NWRs come close to meeting the criterion of an interconnected network of reserves for migratory waterfowl (http://www.refuges.fws.gov/NWRSFiles/Legislation/HR1420/TOC.html). In addition, the FWS's role in protecting seabirds brings it into increased contact with fishery managers over endangered species such as marbled murrelets, kittywakes (salmon fisheries on the West Coast), short-tailed albatross, spectacled eider, and Steller's eiger (in the North Pacific). Consultations under the Endangered Species Act (ESA) focus efforts to develop seabird avoidance and monitoring programs by the fishing fleet and fishery managers.

The legislative history for many NWR areas may "grandfather" in patterns of use that may be seen as destructive or inappropriate for the protection mandate implied in the term "refuge," especially since past practices of management may have been lenient. Refuges usually allow hunting and fishing (recreational and commercial), but there is pressure to modify these practices when damage to resources is demonstrated (Adams, 1993). Public perceptions of refuges are changing such that higher levels of protection are expected, which is pushing management in that direction. There is now a general trend toward more protective management of the NWR System as indicated in the National Wildlife Refuge System Improvement Act of 1997 (P.L. 105-57).

Monitoring and research are a large component of FWS expenditures and personnel activities. Monitoring of wildlife populations, especially wildfowl, takes priority, but "strategic" monitoring of habitat change and other factors affecting wildlife is also undertaken. The reliance of the FWS on field personnel for management and research disperses agency staff over wide areas and into small communities. This has benefits in the form of involvement with stakeholder communities and interests, as well as being able to take enforcement action as necessary when education and deterrence fail.

Fishery Management Areas

Fisheries reserves are defined in Chapter 1 as areas that preclude fishing activity on some or all species to protect essential habitat, rebuild stocks (long-term, but not necessarily permanent, closure), provide insurance against overfishing, or enhance fishery yield. Under current federal practice, these designations are made on the recommendation of regional fishery management councils (Figure 8-2) to the Secretary of Commerce in accordance with the Magnuson-Stevens Fishery Conservation and Management Act (NOAA, 1996a), the principal statute guiding regulation of fisheries in federal waters. The eight regional fishery management councils were established under the Magnuson-Stevens Fishery Conservation and Management Act in an advisory role to NMFS with the responsibility to develop management plans for the fisheries under their jurisdiction. The National Marine Sanctuary Program must work through the regional fishery management councils and NMFS to implement fishery regulations within a sanctuary (see Figure 8-1).

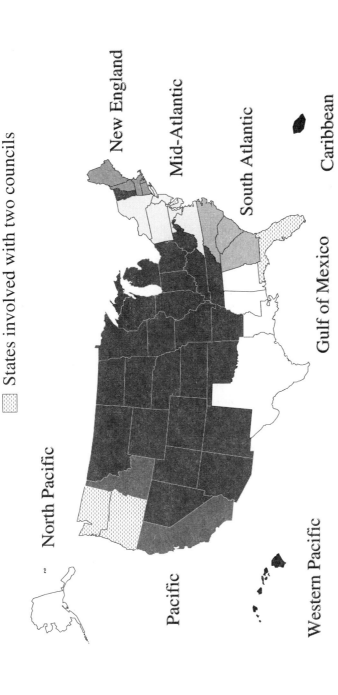

FIGURE 8-2 Map of the eight regional fishery management councils and the states under their jurisdiction. Used with permission of the National Oceanic and Atmospheric Administration.

The fishery management council system represents an innovation in fishery management in the United States through its recognition of regional differences and perspectives in the design of management measures. For many years, public apathy about living marine resources and the intense interests of commercial and recreational interests resulted in fishery management council discussions that were limited in scope, often consisting of acrimonious fights among users. Increasingly, as public awareness of fisheries issues has grown, the public voice is heard more often in the council process. In addition, a variety of other interests now broaden the base of council hearings, comment periods, and committee work (e.g., divers, community groups, tribes, consumers). Concerns are increasingly expressed that council decisionmakers are drawn from conventional user groups that are generally unresponsive to other issues. NMFS involvement in planning and stakeholder processes involving other agencies is occurring with more frequency and in innovative ways.

The use of fishery reserves in fishery management is well recognized in practice as well as in the literature (Murray et al., 1999; NMFS, 1999). Zoning decisions made to restrict fishing gear impacts, reduce bycatch, and protect species during vulnerable life stages are common tools employed in fisheries management (NMFS, 1999). Few no-take reserves exist in federal waters, but there are numerous areas in which partial closures are utilized. For example, between the Gulf of Alaska and the Bering Sea, bottom trawl closures are in place for approximately 81,000 nmi^2, an area about equal in extent to federal waters fished off New England. While the principal purposes of fishery reserves established in the United States are to reduce fishing mortality, protect critical life-history stages, ensure continued fish production, and reduce secondary impacts of fishing (e.g., Alaska: see Ackley and Witherell, 1999; Witherell et al., 2000; New England; see NOAA, 1999; Gulf of Mexico and Southeast Atlantic; see Coleman et al., 1996), these measures may also directly and indirectly confer benefits in the form of conservation of biodiversity and habitat. In some cases, critical habitat is designated for endangered species (e.g., for monk seals and Steller sea lions) but incidentally provides protection for other species. Fisheries, marine mammals, and endangered species management each offer avenues to create reserves and to better manage resources generally.

The effectiveness of fishery reserves has been discussed earlier in this report (Chapters 5 and 6), but there are few examples of reserves that specifically test design principles for optimal fishery enhancement. As pointed out in this study and in others, (e.g., NRC, 1999a), such limited measures as presently in place in most if not all fishery management regions are inadequate to maintain or restore marine fisheries. In some areas, fishery reserves may simply reallocate the catch to recreational rather than commercial fisheries, with no evidence that ecosystem function will improve.

Work by NMFS, and the regional fishery management councils to fully implement the habitat provisions of the 1996 amendments to the Magnuson-

Stevens Fishery Conservation and Management Act (Sections 303(a)[7] and 305(b)), continues and includes serious examination of fishery reserves as a component of the management approach. Fishery reserves presently are being considered by the Pacific Fishery Management Council to protect rockfish (*Sebastes* spp.). The South Atlantic and Gulf of Mexico Fishery Management Councils have a reserve to protect coral habitat known to be important to reef fish. The Gulf of Mexico Fishery Management Council recently approved two marine reserves to evaluate their effectiveness in protecting grouper populations. Under the essential fish habitat provisions of the Magnuson-Stevens Fishery Conservation and Management Act, regional fishery management councils must assess habitat needs of all managed species and amend their fishery management plans accordingly to provide protection where required. In addition, councils are to identify habitat areas of particular concern, and these areas must be given special management attention. To reduce effects of fishing gear on the seabed and avoid bycatch, the North Pacific Fishery Management Council requires that all pollock (*Theragra chalcogramma*) trawling be conducted using midwater trawls. Similarly, the New England Fishery Management Council has closed areas of Georges Bank to bottom trawling and dredging to protect critical habitat for juvenile groundfish (NOAA, 1999).

Because the work on fishery reserves by regional fishery management councils has been driven primarily by single-species concerns and gear conflicts, not by an overall ecosystem-based agenda, these reserves do not constitute an interconnected network. Compliance is generally high and monitoring focuses on stock assessments for the target species. Evaluation of impacts on other components of the ecosystem is neither routine nor at the scale that would allow short- and long-term effects to be detected. In fact, the assessment of fishing effects is generally considered more an area of research than of monitoring. Fully protected reserves would be necessary to provide a baseline against which comparisons can be made with fished areas, in order to distinguish fishing effects from other anthropogenic impacts such as pollution and loss of wetlands (Murray et al., 1999). Fishery managers are optimistic that over the next 5-10 years, declines in fished populations will abate and recoveries of fish stocks will be evident. Others are less sanguine about the pace of change and call for more actions employing the precautionary principle.

Other Federal Areas and Legislation

The U.S. Forest Service (USFS) manages vast areas of public land adjacent to marine waters, primarily in southeast Alaska and California. Its chief legal mandates reside in the Multiple Use Sustained Yield Act 1960, Forest and Rangeland Renewable Resources Planning Act 1974, and Federal Land Policy and Management Act 1976, dealing with managing forests, grazing, water, wildlife, and minerals in terrestrial contexts (Wilkinson and Anderson, 1987; USDA,

1993). It does not appear that federal national forest designations provide authority in intertidal or state waters, although management of uplands and watersheds to protect these areas is an important responsibility (Cubbage et al., 1993). Due to the popularity of fjords and bays surrounded by national forest lands in Alaska at sites such as Misty Fjords and Admiralty Island National Monument (www.alaska.net/~anm), recreational boating and other uses are managed. In northern California, national forestlands and their management are important in controlling runoff of silt that can be a problem in nearshore environments. In Big Sur, California, recent coastal conservancy acquisitions are being added adjacent to national forest lands. Given the limited jurisdiction of the USFS has over marine areas, it is not possible to apply the criteria for evaluation set out in this chapter. It may be expected that the role played by the USFS in the use and protection of adjacent marine waters will increase, especially under President Clinton's executive order on MPAs (Appendix E).

Federal management of military reservations that may restrict or regulate access to shoreline and offshore areas creates de facto MPAs that can be significant for scientific research, fish refugia, and habitat preservation. Two examples are Kaneohe Bay Marine Air Force Base (D. Drigot, personal communication, 1999) and the restricted access zone around the Kennedy Space Center in Florida's Merritt Island National Wildlife Refuge (Johnson et al., 1999). In both of these areas, increased abundance of fish has been associated with restricted access (D. Drigot, personal communication; Johnson et al., 1999).

Besides direct designation of distinct MPAs such as national parks, sanctuaries, and estuarine reserves, existing legislation permits federal agencies to take actions that include establishing no-take reserves and other protective measures that may be equally or more effective depending on the conflict or opportunity at hand (Table 8-3). Other legislation provides ways to mitigate or minimize threats to designated MPAs by virtue of management measures taken outside MPAs. Under existing legislation and the executive order of 2000, federally funded or conducted activities are prevented from harming many key areas such as critical habitats for endangered species. Full and effective use of existing legislative tools could vastly reduce impacts of human disturbance on the marine environment and complement MPA designations as tools for conserving resources or biodiversity.

The Marine Mammal Protection Act (MMPA) (P.L. 92-522) of 1972 authorizes NMFS to take measures to protect marine mammals that may involve setting aside habitat required by various life stages, although the chief provision is the prohibition on "taking" marine mammals directly or indirectly (e.g., through changes in habitat). Similarly, the ESA of 1973 (P.L. 93-205) allows the Secretary of Commerce to identify threatened and endangered species (for most marine fish species, marine mammals, and sea turtles) and to designate habitats critical to their survival. Frequently, combination of the MMPA and the ESA results in designation of critical habitat for species such as Steller sea lion (*Eu-*

TABLE 8-3 Federal Laws Relative to Designation and Management of Marine Protected Areas

Laws That Permit Designation of MPAs	Effect
National Marine Sanctuaries Act. (16 U.S.C. §§ 1431 et seq.)	Designation of National Marine Sanctuaries for the purpose of comprehensive and coordinated conservation and management while facilitating compatible public and private use of resources not prohibited by other authorities
Magnuson-Stevens Fishery Conservation and Management Act (16 U.S.C. §§ 1801-1883)	Establishes fishery management authority over living marine resources on the Continental Shelf and in the EEZ. Includes national standards, regional fishery management councils and requirements for fishery management plans; also includes authority to designate open and closed areas as fishery management tools to protect spawning and rearing populations, essential fish habitat, habitat areas of particular concern, etc.
National Park Service Organic Act (16 U.S.C. §§ 1,2-4)	Creates the National Park Service to administer parks, monuments, and reservations as established by Congress for the purpose of nature preservation and public enjoyment
Coastal Zone Management Act of 1972 as amended (CZMA) (16 U.S.C. §§ 1451 et seq. [See also section 6217 of the Coastal Area Management Act Reauthorization Amendments of 1990 (16 U.S.C. §§ 1455b)]	Establishes the National Estuarine Research Reserve System under which states may seek federal approval for NERRs if the areas qualify as biogeographic and typological representations of estuarine ecosystems that are suitable for long-term research and conservation
National Wildlife Refuge System (16 U.S.C. § 668dd)	Establishes areas for conservation of fish and wildlife, prohibits damage to such resources, and requires a permit for use. Public recreation is permitted to the extent that it is not inconsistent with the primary objectives for which an area is designated
National Wilderness Preservation System 16 U.S.C. § 1131	Congress can designate areas within jurisdictions of land management agencies as part of a system of federal wilderness to be managed so as to preserve its natural conditions

(continued)

TABLE 8-3 *Continued*

Laws That Permit Designation of MPAs	Effect
Laws that Regulate Activities in the Marine Environment	
Oil Pollution Act of 1990, (33 U.S.C. §§ 2701 et seq.)	Improves federal response authority, increases penalties, and requires tank vessel and facility response plans. Defines scope of damages for which there may be liability
Federal Water Pollution Control Act/ Clean Water Act, as amended (33 U.S.C. §§ 1251 et seq.)	Establishes program for restoring and maintaining the chemical, physical, and biological integrity of the waters of the United States, including offshore oil and gas. Section 320 establishes the National Estuary Program, which uses a consensus-based approach for protecting and restoring estuarine ecosystems. Prohibits discharge of oil and hazardous substances into coastal and ocean waters. Requires operable marine sanitation devices (Section 312). Regulates discharge of dredged or fill materials in territorial seas (Section 404)
Ocean Dumping Act (Titles I and II of the Marine Protection, Research and Sanctuaries Act of 1972) (33 U.S.C. §§ 1401 et seq.)	Requires a permit for transportation and dumping of any material in ocean waters (e.g., sewage sludge, industrial wastes, high-level radioactive waste, and medical wastes). Authorizes research and monitoring on pollution impacts
National Environmental Policy Act of 1969 (42 U.S.C. §§ 4321 et seq.)	Requires preparation of a detailed environmental impact statement for every federal action that identifies the potential impacts of such actions and alternative approaches that would mitigate any impacts
Endangered Species Act of 1973 (16 U.S.C. §§ 1531-1543)	Protects species of plants or animals listed as threatened or endangered by requiring the designation of critical habitat and development and the implementation of a recovery plan in such a way as to ensure that any federal action is not likely to jeopardize the existence of the species or its habitat. Prohibits the take of any endangered species
Marine Mammal Protection Act of 1972 (16 U.S.C. §§ 1361-1407)	Determines management responsibilities for protection of cetaceans and pinnipeds, sea otters, polar bears, walruses and manatees

	and establishes a moratorium on taking and importing marine mammals and their products except in special circumstances
Outer Continental Shelf (OCS) Lands Act (43 U.S.C. §§ 1331 et seq.)	Establishes federal jurisdiction over submerged lands in the OCS and allows leasing of minerals and energy resources, management of exploration and development, protection of the marine and coastal environment, development of improved technologies, and opportunities for state and local participation in policy and planning decisions
Archaeological Resources Protection Act of 1979 (16 U.S.C. §§ 470aa et seq.)	Protects historic preservation sites from looting and permits the recovery of these items under permit when inside national parks, wildlife refuges, etc. but does not apply to the OCS

SOURCE: http:www.yoto98.noaa.gov/; U.S. Department of Agriculture, Forest Service, 1993.

metopias jubatus) in central and western Alaska. There, NMFS has banned Alaska pollock fishing within 10-20 miles of rookeries and haulout areas and has severely restricted pollock harvests within the critical habitat of Steller's sea lions.

Additional federal legislation that helps to control human impacts on the marine environment (Table 8-3) includes the Clean Water Act (P.L. 95-217) of 1977; Ocean Dumping Act (Title 1 of the Marine Protection, Research and Sanctuaries Act) of 1972; Coastal Zone Management Act (mentioned earlier re Section 315) of 1972; Coastal Barriers Resources Act (P.L. 97-348) of 1980; Fish and Wildlife Coordination Act (FCWA 48, Stat. 401) of 1934; Marine Plastics Pollution Research and Control Act of 1987; Oil Pollution Act of 1990; Comprehensive Environmental Response, Compensation and Liability Act of 1980 (amended in 1986); National Environmental Policy Act (NEPA, P.L. 91-190) of 1969; and Outer Continental Shelf Lands Act (OCSLA, 67 Stat. 462) of 1953. The Wilderness Act (P.L. 88-577, 16 U.S.C. 1131-1136) is relevant and has potential application in national parks, refuges, and other areas of the marine environment. Because the concept of "marine wilderness" is frequently offered as a type of MPA, it is useful to provide a formal statement of its purpose under the law. The Wilderness Act essentially allows Congress to designate wilderness as "an area where the earth and its community of life are untrammeled by man, where man himself is a visitor who does not remain." The Wilderness Act allows a number of uses in terrestrial ecosystems that appear incompatible with its basic purpose to "preserve its natural conditions" (e.g., grazing and mining).

It also grandfathers in the use of aircraft and motorboats where these uses were preexisting (Keiter, 1988). In terrestrial wilderness areas, nonmechanized recreation is the standard; yet in the marine environment, motorized recreation or access is the norm. However, special legislation may be passed to restrict certain activities, such as the use of jet skis in the Channel Islands. How to define the nature of marine wilderness is a matter of controversy that may take years to resolve.

Obviously, the federal legislative environment in which MPAs are implemented in the United States is complex, with stringent standards for identification and mitigation of environmental impacts, consultative requirements, and pollution prevention and control measures.

State Initiatives

At the state level, the range and number of MPAs are extensive and include NERRs. Recent efforts to catalog MPAs have reported roughly 100 each in the States of Washington and California, including the federal MPAs but not counting proposed sites or intertidal extensions of state parks. Numbers do not tell the whole story. Murray and Ferguson (1998) categorized MPAs into research and educational marine preserves, recreational marine preserves, marine species preserves, marine habitat or nature preserves, and multiple-use protected areas. They found that private entities such as the Nature Conservancy and land trusts, as well as cities and counties, have participated in developing the concept. An innovative approach to no-take fisheries reserves in Washington State is being undertaken by San Juan County with its voluntary bottomfish fish recovery areas (D. Willows, presentation to the National Research Council, September 9, 1999; San Juan County Marine Resources Committee, undated). This community-based model has been accepted and replicated by the Northwest Straits Commission in six other counties in the region (Murray and Metcalf, 1998). In addition, the Washington Department of Natural Resources (1996) is developing management plans for some of its intertidal and subtidal lands that are linked to upland conservation areas.

In California, McArdle (1997) found that although there were 101 MPAs occupying 18.2% of California's state waters, only a small (0.2%) portion of the MPA area was closed to fishing. California places high priority on its coastal and ocean resources and consequently has developed an extensive planning process to better protect them (California Resources Agency, 1997). In 1999, the California legislature passed the Marine Life Protection Act, which will launch the establishment of a network of MPAs and the designation of no-take reserves (MPA News, 1999).

Hawaii is known for the Hanauma Bay Marine Preserve on Oahu, which attracts 1.5 million visitors per year (Moribe, 1999). Less well known is the West Hawaii Regional Fishery Management Area Bill (Act 306). It caps aquarium-collecting permits off the island of Hawaii, designates at least 30% of West Hawaii as off-limits for collecting reef-dwelling species, designates fishery reserves to overlap with previously designated areas, expands a day-use mooring system, and prohibits use of gillnets in some areas. This combination of measures is intended to resolve long-standing conflicts among various user groups, as well as to develop MPAs to restore marine life over significant areas of the coast.

Summary of Current System

MPAs in the United States today are not the result of a systematic effort to design and implement a system of MPAs to serve the multiple goals described in Chapter 2. Existing MPAs are embedded in a matrix of programs and policies under which various jurisdictions (federal, state, local) use their authorities to manage ocean space, activities, and resources. There are comprehensive management programs under federal law for coastal zones, clean water, fisheries, and offshore oil and gas, frequently requiring state and local involvement. In addition to these broad management measures, marine areas have been designated for national parks, estuarine research reserves, marine sanctuaries, and wildlife refuges, as well as fishery reserves. In most cases, state and local programs parallel the legislative efforts at the federal level. Thus, the existing MPAs have evolved out of myriad efforts to protect areas for various purposes and using different tools.

Frequently, the overlapping jurisdictions of various federal and state agencies have presented obstacles to the implementation and management of MPAs, indicating the need for greater coordination among these agencies. There is no federal agency or authority charged with the oversight of marine protected areas. However, Executive Order 13158 of May 26, 2000, may provide a framework for greater coordination among federal agencies (Appendix E). At the state level, California convened an interagency workgroup that developed new, simpler classifications to coordinate the fragmented system of existing marine managed areas, defined as "named, discrete geographic marine and estuarine areas along the California coast designated using legislative, administrative, or voter initiative processes, and intended to protect, conserve or otherwise manage a variety of resources and their uses," and recommended a more inclusive and science-based process for designating new areas (California Resources Agency, 2000). In addition, California's Marine Life Protection Act (AB 993) emphasizes using a more coordinated approach to implement MPAs for marine area management.

Conservation of Biodiversity and Habitat

One could argue, that despite the numerous large and small MPAs found around the United States, marine biodiversity continues to decline. Therefore, more reserves with greater protections should be established. The converse argument is unlikely—that MPAs are part of the problem. However, much of the widespread degradation of the marine environment comes from inadequate regulation of land-based activities that affect the marine environment. Agricultural runoff and other nonpoint sources of pollutants to the Gulf of Mexico ecosystem are implicated in generating the large zone of hypoxic bottom water in the northern Gulf of Mexico (Goolsby, 2000). This is the type of complex management issue addressed under the Clean Water Act. Similarly, coastal hardening affects nearshore habitats and transport of sand. These issues fall, in part, under the aegis of the CZMA. MPAs may not be the complete answer, but their expansion is likely to improve land-based efforts to protect marine biodiversity and habitats.

Fishery Management

The discussion of fishery management above covers primarily federal waters. In state and local areas, there are a large number of fishery reserves established for specific purposes. Unfortunately, relatively little is known about the contribution of these reserves to maintaining or rebuilding fisheries. These areas tend to be small in size and very specific with respect to purpose (e.g., herring spawning area, good habitat for large rockfish). Nearshore environments are often key areas for spawning and rearing of juvenile fish, so their proportional significance in the life histories of some species may be quite large. In addition, it appears that a large proportion of these state and local reserves are proximate to shoreside protected areas as well, and this may show potential for managing across the shoreline boundary. Many of these sites are of relatively recent origin relative to the time scale in which one would expect to see changes in fish populations, species composition, and size distributions. It is also fair to say that few of these have received adequate monitoring or research. It appears that there are considerable opportunities, given the uncertainty in fishery management, to use additional fishery reserves as a hedge against management failure and to learn from what works and what does not. Developing stakeholder processes around a mutual interest of parties in restoring fish stocks can be a key element in this regard where resources are inadequate to fully monitor and perform research.

Other Ecosystem Services

Federally established MPAs clearly demonstrate the provision of a wide variety of ecosystem services. They also illustrate the growing demand for other ecosystem services. State and local designation of MPAs can be responsive to an even wider range of such services, especially where the initiative for the designation derives from local grass-roots, "bottom-up" kinds of requests. These could range from a shoreside community asking for protective designations in its own "front yards," even if they are not necessarily highly ranked sites for fish production or marine biodiversity, to local governments designating "voluntary" zones for fish recovery. As society becomes more aware of the values that it places on the marine environment, the services it provides, and the potential losses due to degradation, it can be expected that managers will be increasingly called upon to protect these services.

9

Conclusions and Recommendations

CONCLUSIONS

Conservation of Marine Resources Requires New Approaches

Globally, there has been a surge of interest in designating areas of the seas as marine reserves and protected areas to maintain and conserve marine species and habitats threatened by human activities. There is a growing consensus that living marine resources require more stringent protections. Crises facing many marine ecosystems are increasing and attracting more public attention. Among these are the recent collapse of the Newfoundland cod fishery, the near-collapse of the groundfish fishery in New England, and the loss of coral reef communities to disease and overfishing. Hence, there is widespread concern among policy-makers, scientists, and the public at large about the current status and uncertain future of marine ecosystems. Better approaches for utilizing and protecting living marine resources are needed; however, choosing the best methods to maintain or restore the health of marine ecosystems is a difficult task for resource managers and a source of disagreement among user groups, scientists, and the conservation community.

One does find general agreement, however, on the shortcomings of most current management policies. For example, conventional fishery management commonly focuses on single species and is concerned with species-specific issues such as maximum sustainable yields, appropriate fishing mortality rates, effects of seasonal closures of certain fishing grounds, and appropriate size limits to protect juveniles and the spawning potential of the stock. Whether or not these single-species management strategies achieve their specific goals, their practice often neglects other important and pervasive problems. Furthermore, regulations designed for one fishery may negatively influence other species on the same fishing grounds through gear conflicts, bycatch, habitat destruction, or subtle but important shifts in predator-prey relationships.

Shortcomings in marine resource management also derive from inadequate coordination among agencies charged with these responsibilities. Frustration rises when conventional approaches fragment management into a myriad of regulations from multiple state and federal agencies, each addressing only one component of the problem. The deficiencies in fishery management and ecosystem protection cannot be overcome by continuation of ocean management on a multijurisdictional basis, in which different species are managed separately, agencies may apply regulations independently of each others, and state and federal policies are not fully coordinated. When this piecemeal approach is followed, the interests of various stakeholders are developed in parallel, some stakeholders receive no representation at all, and instead of integrated management, competing regulations may develop that fail to meet objectives of conservation and sustainable use. Also, this narrowly focused approach to management tends to underrepresent the values of the general public and disproportionately represent organized user groups, whether they are commercial fishers, recreational fishing groups, or dive tour operators.

It is clear that despite good intentions and dedicated effort on the part of resource managers at federal, state, and local levels, most existing strategies to regulate fishing or other removals of living marine resources have neither prevented the decline of these resources nor slowed the destruction of habitat. Increasingly, methods are being sought that preserve ecosystem components essential for the health of marine resources, especially when such overarching factors as genetic diversity, species diversity, spawning biomass, and ecosystem stability require protection. Thus, new approaches are necessary to allow a more integrated and comprehensive attack on problems that transcend the concerns of single-species management.

Marine Reserves and Protected Areas Provide a Strategy for Ecosystem-Based Management

A growing body of literature documents the effectiveness of marine reserves for conserving habitats, fostering the recovery of overexploited species, and maintaining marine communities. There is a rising demand for ecosystem-based approaches to marine management that consider the system as a whole rather than as separable pieces of an interlocking puzzle. Congress recognized this in the 1996 reauthorization of the Magnuson-Stevens Fishery Conservation and Management Act (NOAA, 1996a) and requested that the National

Marine Fisheries Service undertake a study of what such management would entail from a fishery management perspective. The study produced a list of key policy objectives "to change the burden of proof, apply the precautionary approach, purchase insurance against unforeseen adverse ecosystem impacts, learn from management experiences, make local incentives compatible with global goals, and promote participation, fairness, and equity in policy and management." The report recommends that fishery management councils "use a zone-based management approach to designate geographic areas for prescribed uses. Such zones could include marine protected areas (MPAs), areas particularly sensitive to gear impacts, and areas where fishing is known to negatively affect the trophic food web" (NMFS, 1999).

Networks of marine reserves, where the goal is to protect all components of the ecosystem through spatially defined closures, should be included as an essential element of ecosystem-based management. Incorporation of MPAs, including marine reserves, into a broader plan for coastal and ocean management offers an opportunity to revise current fragmented management approaches and provide for more inclusive representation of stakeholders concerned about the health of marine ecosystems. The performance of MPAs as a conservation tool is best viewed in the broadest context of management objectives that encompass the full range of human interests in the sea. In this sense, management for direct use (e.g., fisheries), for indirect use (e.g., heritage and existence values), and for ensuring protection of essential ecosystem services ultimately must be accomplished through zoning, which requires designating different areas to meet different goals.

RECOMMENDATIONS

Designing Marine Protected Areas

The design of MPAs should proceed through four stages: (1) evaluate needs, (2) set goals, (3) assemble data on the region to be served by the MPAs, and (4) outline various options for siting areas that meet the previously agreed-on goals. Each stage of this process should involve the broad community of stakeholders—users, managers, scientists, conservationists, homeowners, and other concerned members of the public—such that the final MPA plan represents a joint effort between affected communities and management agencies.

Establishment of networks of marine reserves and protected areas will provide an ecosystem-based approach for meeting the multiple objectives of coastal and marine area management. These objectives include protection of habitat, biodiversity, and fisheries and promotion of research to increase the effectiveness of various conservation and management measures such as empirical determinations of fishing and natural mortality rates to improve the accuracy of stock assessment methods used for fishery management.

To Protect Biodiversity

In the design of a system of marine reserves and protected areas, the complete spectrum of habitats supporting marine biodiversity should be included with emphasis on safeguarding ecosystem processes. One of the best-supported goals of MPAs is to conserve and restore marine biodiversity—that is, to maintain species diversity and the natural balance of species interactions. This goal entails (1) setting aside representative areas of each different habitat in a biogeographic region, (2) establishing systems of marine reserves that are interconnected and large enough to be mostly self-sustaining, and (3) including each habitat type in multiple reserves to provide buffers against changing environmental and societal forces. Connectivity among reserves should be a factor in the design of MPA networks to prevent genetic isolation of populations and to ensure that dispersal of early life stages and re-colonization are facilitated. Moreover, properly networked MPAs will promote habitat linkages necessary for various life stages and ensure continuity of life processes within the MPA network.

To Improve Fisheries Management

Another prominent goal of marine reserves and protected areas is to make fishery management more effective. Fishery reserves can be used to address one or more of the following objectives, dependent on the needs of the fishery:

1. *Allow depleted fisheries to recover from overfishing.* Establishment of reserves to restore depleted fisheries will generally show the largest increases in abundance, size, and age structure of fish stocks within their boundaries, compared to unprotected areas. Fishers are more likely to accept reserves as an alternative management option when stocks are severely overexploited.

2. *Prevent the collapse of fish stocks.* Appropriately designed and implemented reserves can help to prevent severe overexploitation of some fishery resources. Reserves that enclose critical habitat (e.g., nursery or spawning habitats) will be most effective in promoting these fishery management goals.

3. *Improve sustainable yield of fisheries.* For some stocks, the spillover of juveniles and adults from reserves to fishing grounds has the potential to enhance the long-term yield of the fishery. However, evaluation of age- and size-specific dispersal probabilities is essential and critical to understand this enhancement potential. Some reserves may serve as source areas, replenishing depleted or heavily fished stocks in areas that remain open to fishing. In this case, the reserve must be large enough to protect the population within its boundaries, but not so large that dispersal to the surrounding area is limited. Similarly, the reserve should be sited such that oceanographic features provide both sufficient productivity to support protected populations and favorable currents to facilitate appropriate adult and larval spillover to open areas and other reserves. An area designated as an ecological reserve may not export all species of fish to

the surrounding area but may still benefit these fisheries indirectly by supporting biodiversity and other ecosystem services. However, even an effective reserve design may not overcome the danger that there will be increased fishing pressure on unprotected populations in the open areas.

4. *Reduce bycatch of nontargeted species and undersized individuals of targeted species.* Properly designed reserves can protect small fish and other species that otherwise would be caught and killed unintentionally. Frequently, bycatch includes young individuals of a target species—this reduces the size and productivity of the population available to the fishery in subsequent years. Such protection will not only improve the productivity of targeted species but also help maintain the structural and functional integrity of marine communities. Likewise, reserves can help protect threatened or endangered species (e.g., marine mammals and seabirds) either directly by reducing bycatch or indirectly by protecting their food sources.

Marine reserves may provide the only effective means to ensure against overfishing of some species if exploitation is high and there is substantial uncertainty in the stock assessments. Empirical and modeling studies demonstrate that reserves are effective in increasing the population density, biomass, and age structure of species that have limited adult mobility such as benthic invertebrates and some demersal fishes. Conventional management of such species often fails because they have life-history characteristics that are difficult to evaluate using standard stock assessment methods. As an example, rockfishes (*Sebastes* spp.), in addition to low adult mobility, have the following life-history characteristics: extreme longevity, low natural mortality, infrequent recruitment success, low productivity, habitat specificity, and co-occurring species that are fished as an assemblage. In such fisheries, large reserves may prove the most effective means of regulating fishing because a portion of the population is protected independently of the accuracy of the stock assessment.

Modeling studies on several fish species indicate that reserves would have to be extremely large if they were the only or the primary means to maintain fishing rates within sustainable levels. These models predict that the open area should encompass the fraction of the population that could be sustainably fished. Smaller area closures may be sufficient when used in concert with more conventional means of limiting fishing effort. The question of how much fishing ground to close will thus depend on the effectiveness of fishery regulations in open areas. Because of the diversity of fish species and management objectives, it is impossible to set a universal percentage for area closure. Models that incorporate species-specific parameters have derived values ranging from 10% to 80% closure of fishing grounds. Future modeling studies should evaluate the relative performance of a suite of management strategies, including reserves, to regulate landings and fishing mortality rates. These studies should be conducted for specific fisheries, using detailed models to represent the spatial dynamics of populations and fishing effort, while incorporating realistic assumptions about the performance of other methods to control fishing.

To Balance Costs and Benefits

When comparing marine reserves to conventional management, the contribution of reserves to the conservation of biodiversity should be included in the assessment of the costs and benefits of regulating fisheries. In some circumstances it is possible that reserves will yield both classes of benefits simultaneously, but in other cases there will be trade-offs and conflicts between fishery management and conservation of biodiversity. Where trade-offs and conflicts exist, policymaking requires assessment, forecasting, and analysis of how reserves will affect both biodiversity and fisheries. Increased research on the valuation of the full range of potential costs and benefits will be needed to accurately assess the economic impacts of marine reserves.

Whenever benefits of a management approach are discussed, measurements of success must take into account the time paths of benefits and costs as well as the possible changes in the distribution of benefits and costs among different members of the community over time. For instance, reduced fishing access with institution of an ecological reserve may represent an immediate cost to fishers—a cost remaining in play until, and if, the fishery in question regenerates from the protective effects of the reserve, whereas members of the community supported by ecotourism may benefit immediately.

If benefits are defined strictly in terms of market values, the following circumstances might lead to MPAs and reserves providing no net economic gain:

1. where financially profitable activities are displaced by valuable human uses that do not generate financial profits,

2. where the absence of resource security generates incentives to engage in destructive short-term activities rather than sustainable long-term use of marine natural resources,

3. where the costs of enforcement exceed the financial benefits, and

4. where displacement of fishing effort (or other activities) results in a disproportionate degradation of surrounding areas.

In any analysis of costs and benefits, however, it is important to include the nonmarket benefits of preserving the naturalness of ocean areas.

To Complement Other Management Efforts

In designing MPAs and reserves for conserving biodiversity and managing fish stocks, it is important to recognize that the goal is to maintain the health of marine ecosystems beyond the relatively small area protected with-

in reserves. MPA networks can provide a key component of precautionary management frameworks to help secure the long-term persistence of species and ecosystem processes, as well as offer support to fisheries. However, even with closures of representative areas within a given biogeographic zone, the surrounding area will remain vulnerable to the destructive aspects of human activities. Realistically, core ecological reserves will have to be supplemented by additional measures, including conventional fishery regulations in open areas as well as prohibitions on damaging activities that have been poorly regulated in the past.

To Protect a Sufficient Fraction of Marine Habitats

The amount of area needed in MPAs and reserves to preserve ecosystem functioning will depend upon the effectiveness of resource management and environmental regulations both within and outside the MPAs. The complexities and uncertainties of managing living marine resources have fueled the debate over what fraction of the ocean should be fully protected in marine reserves. There has been much discussion of the need for closure of 20% of the seas, sometimes expressed as 20% of the management area, 20% of the exclusive economic zone (EEZ), or 20% of each major ecosystem. This figure was derived initially from a fishery stock assessment model used to determine the level of spawning biomass[1] required to sustain the stock and augmented by attempts to estimate the amount of habitat needed to support fisheries and biodiversity. But, the primary consideration for implementing marine reserves should be the needs of each biogeographical region based on protecting critical habitats (such as spawning grounds, nursery grounds, or other areas harboring vulnerable life stages) and special features (such as seamounts, hydrothermal vents, and coral reefs). The extent of a particular area that needs protection will be determined by the quality and amount of habitat, the current health of the living resources, the efficacy of other management tools, and the rarity of the species to be protected.

An incremental approach to implementing MPAs and reserves should be adopted to protect the areas with the highest conservation needs and greatest ecosystem impact first, with additional areas added as necessary to meet management goals. Although protection of 20% of a management area may be suitable for some circumstances, the fraction of area required will vary by region and management goal. The primary emphasis should be on protection of valuable and vulnerable areas, rather than on achievement of a percentage goal for any given region.

---

[1] Spawning biomass refers to the biomass of mature females.

Implementing Marine Reserves and Protected Areas

Integrating Habitat and Resource Protection with Human Needs and Values

Choosing a location for a marine reserve or protected area requires an understanding of probable socioeconomic impacts as well as the environmental criteria for siting. The suite of studies, recommendations, and management policies involved in the design of reserves for protection of biodiversity and fishery stocks must be tightly integrated with socioeconomic analysis. Adequate consideration of socioeconomic issues will be essential for building public support for MPAs and reserves, without which they will be difficult to implement and enforce.

The economic health of coastal communities depends inherently on the health of the waters that border them. A robust marine ecosystem is valuable not only for fishing, but also for recreation, tourism, and the natural scenic attractions that make the seashore such a desirable place to live and vacation. Thus, in many cases the design of MPAs must consider the socioeconomic issues entailed in coastal land use, issues that loom with increasing importance as larger numbers of people elect to live on our coastlines.

It is essential to involve all potential stakeholders at the outset to develop plans for MPAs that enlist the support of the community and serve local conservation needs. A fundamental lesson learned from experience throughout the world is that attempts to implement MPAs in the absence of general community support invariably fail. Inclusion of "bottom-up" or "grass-roots" approaches to planning, design, and implementation of MPAs offers the best opportunity to develop plans with the endorsement of local communities. **Three major steps are necessary for effective public participation: (1) identify all stakeholders; (2) assess the needs and concerns of the affected communities; and (3) involve stakeholders in MPA planning, design, and implementation.** When communities have a voice and are incorporated into the MPA planning process, they are more likely to share critical knowledge and develop an interest in the long-term success of the MPA.

When identifying stakeholders, it is important to recognize that there are many people in addition to traditional user groups who take an interest in the health of the marine environment. Both on-site and off-site constituents value aesthetics, biodiversity, and conservation. Their views and values should be assessed for prospective MPA locations that include different types of cultural and commercial resources, such as coral reefs, kelp forests, whale habitats, fish spawning sites, or other habitats critical for the survival and productivity of marine species. Such evaluations ought to be conducted in an interdisciplinary setting, with marine scientists, social scientists, survey design specialists, and valuation economists. Results of this type of research would help to properly weigh the public good benefits of MPAs, benefits that tend to be underrepresented in a political process often dominated by more readily quantified user benefits and costs, such as fisheries.

Systematic social studies should be conducted to accurately evaluate the impacts of a proposed MPA on community stability. The overall impact of an MPA on a community may be positive or negative or may simply involve shifts of resources from one community sector to another. Because the impact of an MPA will change over time, the effects on different communities will require assessment over many years. Impact assessment will require analysis of multigenerational attitudes, rather than "snapshot" surveys, to determine the cultural commitments to marine areas.

Monitoring MPA and Reserve Performance

Marine reserves and protected areas must be monitored and evaluated to determine if goals are being met and to provide information for refining the design of current and future MPAs and reserves. As in other resource management situations, the ability to adapt or modify existing MPAs is important to optimize benefits from this management tool. Monitoring and evaluation will also contribute to our basic understanding of marine ecosystems. Indeed, in many cases, marine reserves especially will facilitate important experiments in marine ecology, often at spatial and temporal scales that are unusual in ecological research.

The basic knowledge gained from marine reserves about structure, function, and variability in marine ecosystems will enhance our abilities to design reserves and allow more accurate evaluations of their ecological and socioeconomic consequences. Reserves also allow more accurate estimation of parameters such as natural mortality rates—an essential variable in stock assessment models used to manage fisheries. Without such information, assessments may prove unreliable; hence, regulations based on these assessments are more likely to be challenged either by the fishing industry or by conservationists.

Research in marine reserves and protected areas will help answer questions such as the following with regard to both ecological and socioeconomic performance:

1. What are the impacts of closures on species within reserves and in adjacent open areas?

2. What are the dispersal potentials for adults and early life stages of commercially or ecologically significant species?

3. How do differences in reserve boundary or area ratios affect dispersal?

4. How do oceanographic features influence reserve performance?

5. What are the relative impacts of natural variability and human activities on ecosystem functioning (using comparative studies of reserves and areas open to human use and exploitation)?

6. How do closures affect various user groups, including the interests of future generations?

7. Do attitudes and compliance change over time in communities bordering reserves?

Fishery and resource managers should develop and implement management policies that place more emphasis on spatial approaches to experimentally "explore" their systems and to increase our understanding of how fishing impacts the ecosystem. Much of the research described above will further our understanding of the spatial dimension of fishing and enhance our use of spatial management tools, including rotating fishing zones, experimental policy zones, temporary recovery areas, and spatially limited entry licenses and quotas. Such tools have been applied sparingly in marine fishery management, but they are likely to be valuable approaches for controlling effort and fishing mortality. Also, spatial tools have the potential to allow simultaneous comparisons of different regulatory policies using zoning to delineate replicated management areas. This is necessary to account for the interannual variability in conditions that also affect resources. Thus, instead of relying on uniform, fishery-wide, steady-state policies and statistics, managers and fishery scientists could separate areas for different fishing "treatments" without necessarily reducing the targeted yield for the fishery. In this model, reserves serve as controls to provide baseline values for management experiments.

Also, fishery managers should develop data-gathering systems that enable finer spatial analysis of exploited ecosystems. Many fishery management systems now utilize aggregate and fishery-wide data gathering. Increasing the emphasis on spatial regulations will require more consideration of the location of fishing activity and more spatially oriented stock assessment tools. Modern global positioning system (GPS) technology, vessel monitoring systems (VMSs), and data entry systems are in place in many fisheries already. The use of these tools should be encouraged and expanded, with the aim of increasing the understanding of the spatial and temporal distribution of fishing activity and yield.

Supporting MPAs Through Institutional Coordination

The design, implementation, and monitoring of MPAs and reserves require effective institutional structures at federal, state, and local levels of management. Regional coordination of management will be required to establish networks of MPAs and to designate zones for specific uses. As emphasized earlier, fragmented management policies may result in different agencies working at cross-purposes. Hence, integration of management among agencies is essential, and current programs should work together to develop a policy on MPAs. In developing this policy, agencies should recognize all groups with strong interests in the sea, ensuring opportunities for input from those concerned with biological diversity, ecosystem functioning, and the protection of the nation's marine heritage. At the federal level, these agencies would include the National Marine Fisheries Service, National Marine Sanctuaries, and National Estuarine Research Reserves (National Oceanic and Atmospheric Administration), the National Estuary Program (Environmental Protection Agency), the Fish and Wildlife Service, and the National Park Service (Department of the Interior) established under federal legislation, and equivalent agencies at the level of the states and territories. Existing programs should be used both as starting points in establishing a comprehensive system of MPAs and as vehicles for monitoring and evaluating the impacts of MPAs and reserves on fisheries, biodiversity, and ecosystem functioning.

The potential economic and ecological benefits of marine reserves and protected areas will not be realized without a sufficient commitment to enforcement and monitoring. To maintain programmatic integrity and public support for marine reserves and protected areas, it is necessary that sufficient regulatory authority and funds for enforcement, research, and monitoring be provided to implement management plans. Effective enforcement of reserves is essential to sustain cooperation from the general public and affected user groups. Upgraded monitoring programs will ensure that robust data are collected for application both locally and regionally. Results from monitoring programs should be integrated with research programs for the evaluation of reserve performance and design of more effective marine reserves.

MPAs should be developed as a component of broader coastal zone and continental shelf management. This approach represents a shift toward more spatially explicit management of marine resources in recognition of the need to protect areas representative of the complete range of marine species and habitats. Finally, the management system should be adaptable so that the knowledge gained from research and monitoring can be applied to improve performance through more effective design.

Sufficient scientific information exists on the habitat requirements and life-history traits of many species to support the implementation of marine reserves and protected areas to improve management. Given the complexity of natural ecosystems and the broad range of conservation objectives, there are bound to be uncertainties about the optimal design of MPAs, particularly when they include ecological and fishery reserves. However, these uncertainties should not be used as an excuse for failing to take action. Even the most thorough studies of MPAs and reserves, or other management tools, will not eliminate uncertainties with respect to performance. Rather, optimization in the design of MPAs and reserves will depend on an iterative process that combines careful planning with experience. Prevention of the continuing erosion of quality in the marine environment is a shared global concern that requires fresh approaches to management, including MPAs. MPAs, including marine reserves, should be more broadly implemented to improve management of the marine environment and ensure that future generations will benefit from the ocean's bounty.

References

Abramson, P.R., and R.F. Inglehart. 1995. *Value Change in Global Perspective*. University of Michigan Press, Ann Arbor.

Ackley, D., and D. Witherell. 1999. Development of a marine habitat protection area in Bristol Bay, Alaska. Pp. 511-526 in *Alaska Sea Grant College Program: Ecosystem Approaches for Fisheries Management*. AK-SG-99-01. Alaska Sea Grant College Program, Fairbanks.

Adams, D. 1993. *Renewable Resource Policy: The Legal-Institutional Foundations*. Island Press, Washington, D.C.

Adlerstein, S.A., and R.J. Trumble. 1998. Size-specific dynamics of Pacific halibut: a key to reduce bycatch in the groundfish fishery. *International Pacific Halibut Commission Technical Report* No. 39.

Agardy, T. 1994. Advances in marine conservation: The role of marine protected areas. *Trends in Ecology and Evolution* 9:267-270.

Agardy, T.S. 1997. *Marine Protected Areas and Ocean Conservation*. Academic Press and R.G. Landes Co., Georgetown, Texas.

Agardy, T.S. 2000. Effects of fisheries on marine ecosystems: A conservationist's perspective. *ICES Journal of Marine Science* 57(3):761-765.

Alcala, A.C. 1988. Effects of marine reserves on coral fish abundances and yields of Philippine coral reefs. *Ambio* 17:194-199.

Alcala, A.C., and G.R. Russ. 1990. A direct test of the effects of protective management on abundance and yield of tropical marine resources. *Journal du Conseil* 47(1):40-47.

Alcala, A.C., and A.T. White. 1984. Options for management. Pp. in R.A. Kenchington and B.E.T. Hudson (eds.), *Coral Reef Management Handbook*. UNESCO, Jakarta, Indonesia.

Allison, G., J. Lubchenco, and M. Carr. 1998. Marine reserves are necessary but not sufficient for marine conservation. *Ecological Applications* 8(1):S79-S92.

Allison, G.W., S. Gaines, J. Lubchenco, and H. Possingham. In review. Ensuring persistence of marine reserves: Catastrophes require adopting an insurance factor. *Ecological Applications*.

Alverson, D.L., M.H. Freeberg, S.A. Murawski, and J.G. Pope. 1994. *A Global Assessment of Fisheries Bycatch and Discards.* FAO Fisheries Technical Paper 339. Food and Agriculture Organization, Rome.

Ames, E.P. 1998. *Cod and haddock spawning grounds in the Gulf of Maine.* Island Institute, Rockland, Maine.

Anderson, E.D., R.K. Mayo, K. Sosebee, M. Terceiro, and S.E. Wigley. 1999. Northeast demersal fisheries. Pp. 89-97 in *Our Living Oceans: A Report on the Status of U.S. Living Marine Resources, 1999.* U.S. Government Printing Office, Washington, D.C.

Attwood, C.G., and B.A. Bennett. 1994. Variation in dispersal of galjoen (*Coracinus capensis*) (Teleostei: Coracinidae) from a marine reserve. *Canadian Journal of Fisheries and Aquatic Science* 51:1247-1257.

Attwood, C.G., and B.A. Bennett. 1995. Modeling the effect of marine reserves on the recreational shore-fishery of the south-western cape, South Africa. *South African Journal of Marine Science* 16:227-240.

Attwood, C.G., J.M. Harris, and A.J. Williams. 1997. International experience of marine protected areas and their relevance to South Africa. *South African Journal of Marine Science* 18:311-332.

Ault, J.S., J.A. Bohnsack, and G.A. Meester. 1998. A retrospective (1979-1996) multispecies assessment of coral reef fish stocks in the Florida Keys. *Fishery Bulletin* 96(3):395-414.

Babcock, R.C., S. Kelly, N.T. Shears, J.W. Walker, and T.J. Willis. 1999. Large-scale habitat change in temperate marine reserves. *Marine Ecology Progress Series* 189:125-134.

Baker, A. 2000. Supreme Court to consider Glacier Bay appeal. *Pacific Fishing* 21(8):16.

Ballantine, W.J. 1991. *Marine Reserves for New Zealand.* Leigh Laboratory Bulletin 25. University of Auckland, New Zealand.

Ballantine, W.J. 1997. Design principles for systems of "no-take" marine reserves. *Workshop on the Design and Monitoring of Marine Reserves, February 18-20.* Fisheries Centre, University of British Columbia, Canada.

Bannerot, S.P. 1984. *The dynamics of exploited groupers (Serranidae): An investigation of the protogynous hermaphroditic reproductive strategy.* Ph. D. dissertation. University of Miami, Coral Gables, Fla.

Barber, P., S.R. Palumbi, M. Erdmann, and K. Moosa. 2000. A marine Wallace's line. *Nature* 406: 692-693.

Barry, J.P., C.H. Baxter, R.D. Sagarin, and S.E. Gilman. 1995. Climate-related, long-term faunal changes in a California rocky intertidal community. *Science* 267:198-201.

Baumol, W., and W. Oates. 1988. *The Theory of Environmental Policy.* Cambridge University Press, Cambridge, U.K.

Bayle-Sempere, J.T., A.A. Ramos-Espla, and J.A. Garcia Charton. 1994. Intra-annual variability of an artificial reef fish assemblage in the marine reserve of Tabarca (Alicante, Spain, SW Mediterranean). *Bulletin of Marine Science* 55:824-835.

Beets, J., and A. Friedlander. 1992. Stock analysis and management strategies for the red hind, *Epinephelus guttatus*, in the U.S. Virgin Islands. Pp. 66-79 in *Proceedings of the 42nd Annual Meeting of the Gulf and Caribbean Fisheries Institute, Jamaica, 1989.*

Beets, J., and A. Friedlander, 1999. Evaluation of a conservation strategy: A spawning aggregation closure for red hind, *Epinephelu guttatus*, in the U.S. Virgin Islands. *Environmental Biology of Fishes* 55:91-98.

Bell, J.D. 1983. Effects of depth and marine reserve fishing restrictions on the structure of a rocky reef fish assemblage in the north-western Mediterranean Sea. *Journal of Applied Ecology* 20:357-369.

Bennett, B.A., and C.G. Attwood. 1991. Evidence for recovery of a surf-zone fish assemblage following the establishment of a marine reserve on the southern coast of South Africa. *Marine Ecology Progress Series* 75(2-3):173-181.

Bennett, B.A., and C.G. Attwood. 1993. Shore-angling catches in the De Hoop Nature Reserve, South Africa, and further evidence for the protective value of marine reserves. *South African Journal of Marine Science* 13:213-222.

Bennett, J.W. 1984. Using direct questioning to value the existence benefits of preserved natural areas. *Australian Journal of Agricultural Economics* 28(2, 3):136-152.

Bertness, M., and S. Gaines. 1993. Larval dispersal and local adaptation in acorn barnacles. *Evolution* 47(1):316.

Beverton, R.J.H., and S.J. Holt. 1957. *On the Dynamics of Exploited Fish Populations.* Chapman & Hall, New York. 533 p.

Bingham, B.L. 1992. Life histories in an epifaunal community—coupling of adult and larval processes. *Ecology* 73:(6) 2244-2259.

Black, K.P. 1993. The relative importance of local retention and inter-reef dispersal of neutrally buoyant material on coral reefs. *Coral Reefs* 12(1):43.

Block, B.A., H. Dewar, T. Williams, E.D. Prince, C.D. Farwell, and D. Fudge. 1998a. A new satellite technology for tracking the movements of Atlantic bluefin tuna. *Proceedings of the National Academy of Sciences USA* 95:9384-9389.

Block, B.A., H. Dewar, T. Williams, E. Prince, and C.D. Farwell. 1998b. The use of archival, acoustic, and satellite tags on tuna. *Marine Technological Society Journal* 32:37-46.

Bockstael, C.F., N.M. Hanemann, and C. Kling. 1987. Estimating the value of water quality improvements. *Water Resources Research* 23:951-960.

Boehlert, G.W. 1996. Larval dispersal and survival in tropical reef fishes. Pp. 61-84 in N.V.C. Polunin and C.M. Roberts (eds.), *Reef Fisheries.* Chapman and Hall, London.

Bohnsack, J.A. 1990. How marine fishery reserves can improve reef fisheries. Pp. 217-241 in M.H. Goodwin, G.T. Waugh, and M.J. Butler (eds.), *Proceedings of the 43rd Gulf and Caribbean Fisheries Institute.* Gulf and Caribbean Fisheries Institute, Charleston, S.C.

Bohnsack, J.A. 1992. Reef resource protection: The forgotten factor. *Marine Recreational Fisheries* 14:117-129.

Bohnsack, J.A. 1996. Maintenance and recovery of reef fishery productivity. Pp. 283-313 in N.V.C. Polunin and C.M. Roberts (eds.), *Reef Fisheries.* Chapman and Hall, London.

Bohnsack, J.A. 1998. Application of marine reserves to reef fisheries. *Australian Journal of Ecology* 23:298-304.

Botsford, L.W., J.C. Castilla, and C.H. Pearson. 1997. The management of fisheries and marine ecosystems. *Science* 277:509-515.

Botsford, L.W., L.E. Morgan, D.R. Lockwood, and J.E. Wilen. 1999. Marine reserves and management of the northern California red sea urchin fishery. *California Cooperative Oceanic Fisheries Investigations Reports* 40:87-93.

Braden, J., and C. Kolstad. 1991. *Measuring the Demand for Environmental Quality.* Elsevier Science, Amersterdam.

Bradbury, A. 1990. Management of the commercial dive fisheries for sea urchins and sea cucumbers. Pp. 56-65 in J.W. Armstrong and A.E. Copping (eds.), *Status and Management of Puget Sound's Biological Resources.* EPA 910/9-90-001. U.S. Environmental Protection Agency. Seattle, WA.

Bradbury, A. 1991. Management and stock assessment of the red sea urchin (*Strongylocentrotus franciscanus*) in Washington State: Periodic rotation of the fishing grounds. *Journal of Shellfish Research* 10:233 (abstract only).

Brander, K. 1981. Disappearance of common skate *Raja batis* from the Irish Sea. *Nature* 290(5801):48-49.

Brookshire, D., B. Ives, and W. Schulze. 1976. The valuation of aesthetic preferences. *Journal of Environmental Economics and Management* 3(4):325-346.

Brown, B. 1997. Coral bleaching: Causes and consequences. *Proceedings from the 8th International Coral Reef Symposium* 1:65-74.

Brown, G., and R. Mendelsohn. 1984. The hedonic travel cost model. *Review of Economics and Statistics* 66:427-433.

Bustamante, R.H., and J.C. Castilla. 1990. Impact of human exploitation on populations of the intertidal southern bullkelp *Durvillae antarctica* (Phaeophyta, Durvilleales) in central Chile. *Biological Conservation* 52:205-220.

Bustamante, R.H., G.M. Branch, and S. Eeekhout. 1995. Maintenance of an exceptional grazer biomass on South African intertidal shores: Trophic subsidy by subtidal kelps. *Ecology* 76:2314-2329.

Bustamante, R.H., P. Martinez, F. Rivera, R. Bensted-Smith, and L. Vinueza. 1999. A proposal for the initial zoning scheme of the Galapagos Marine Reserve. *Charles Darwin Research Station Technical Report*, October.

Buxton, C.D. 1993. Marine reserves—The way ahead. Pp. 170-174 in *Fish, Fishers, and Fisheries: Proceedings of the 2nd South African Marine Linefish Symposium, 23-24 Oct 1992*, L.E. Beckley and R.P. Van der Elst, eds., Oceanographic Research Institute, Durban.

Buxton, C.D., and M.J. Smale. 1989. Abundance and distribution patterns of three temperate marine reef fish (Teleostei: Sparidae) in exploited and unexploited areas off the southern Cape coast. *Journal of Applied Ecology* 26:441-451.

Caddy, J.F. 2000. Marine catchment basin effects versus impacts of fisheries on semi-enclosed seas. *ICES Journal of Marine Science* 57:628-640.

California Resources Agency. 1997. *California's Ocean Resources: An Agenda for the Future.* California Resources Agency, Sacramento.

California Resources Agency. 2000. *Improving California's System of Marine Managed Areas: Final Report of the State Interagency Marine Managed Areas Workgroup.* California Resources Agency, Sacramento.

Camber, C.I. 1955. A survey of the red snapper fishery of the Gulf of Mexico, with special reference to the Campeche Banks. *Florida Board on Conservation Technical Series No. 12.* Marine Laboratory, University of Miami, Coral Gables, FL.

Campana, S.E. 1999. Chemistry and composition of fish otoliths: Pathways, mechanisms and applications. *Marine Ecology Progress Series* 188:263-297.

Campana, S.E., J.A. Gagné, and J.W. McLaren. 1995. Elemental fingerprinting of fish otoliths using ID-ICPMS. *Marine Ecology Progress Series* 122:115-120.

Campana, S.E., G.A. Chouinard, J.M. Hanson, A. Frechet, and J. Brattey. 2000. Otolith elemental fingerprints as biological tracers of fish stocks. *Fisheries Research* 46:343-357.

Carlton, J.T. 1992. Dispersal of living organisms into aquatic ecosystems as mediated by aquaculture and fisheries activities. Pp. 13-45 in A. Rosenfield and Roger Mann (eds.), *Dispersal of Living Organisms into Aquatic Ecosystems.* Maryland Sea Grant Publication, College Park.

Carlton, J.T. 1993. Neoextinctions of marine invertebrates. *American Zoologist* 33:499-509.

Carr, M.H., and P.T. Raimondi. 1998. Concepts relevant to the design and evaluation of fishery reserves. Pp. 27-31 in M.M. Yoklavich (ed.), *Marine Harvest Refugia for West Coast Rockfish: A Workshop.* NOAA Technical Memorandum. NOAA-TM-NMFS-SWFSC-225. U.S. Department of Commerce, Pacific Grove, CA.

Carr, M.H., and D.C. Reed. 1993. Conceptual issues relevant to marine harvest refuges: Examples from temperate reef fishes. *Canadian Journal of Fisheries and Aquatic Science* 50:2019-2028.

Carson, R.T., R.C. Mitchell, W.M. Hanemann, S. Presser, R.J. Kopp, and P.A. Ruud. 1992. A contingent valuation study of lost passive use values resulting from the *Exxon Valdez* oil spill, a report to the Attorney General of the State of Alaska, November 10, 1992.

Carlton, J.T., , D.P. Cheney, and G.J. Vermeij. 1982. Ecological effects and biogeography of an introduced marine species: The periwinkle, *Littorina littorea. Malacological Review* 15:143-150.

Carter, J., and G.R. Sedberry. 1997. The design, function and use of marine fishery reserves as tools for the management and conservation of the Belize barrier reef. *Proceedings of the 8th International Coral Reef Symposium* 2:1911-1916.

Casey, J.M., and R.A. Myers. 1998. Near extinction of a large widely distributed fish. *Science* 281:690-692.

Castilla, J.C. 1993. Humans: Capstone strong actors in the past and present coastal ecology play. Pp. 158-162 in M.I. McDonnell and S.T.A Pickett (eds.), *Humans as Components of Ecosystems.* Springer-Verlag, New York.

Castilla, J.C. 1999. Coastal marine communities: Trends and perspectives from human exclusion experiments. *Trends in Ecology and Evolution* 14:280-283.

Castilla, J.C., and L.R. Durán. 1985. Human exclusion from the rocky intertidal zone of southern Chile: The effects on *Concholepas concholepas* (Gastropoda). *Oikos* 45:391-399.

Castilla, J.C., and M.A. Varas. 1998. A plankton trap for exposed rocky intertidal shores. *Marine Biology Progress Series* 175:299-305.

Chadwick, D.H. 2000. The Samoan way. *National Geographic* 198(1):72-89.

Chapin, F.S. III, E.S. Zavaleta, V.T. Eviner, R.L. Naylor, P.M. Vitousek, H.L. Reynolds, D.U. Hooper, S. Lavorel, O.E. Sala, S.E. Hobbie, M.C. Mack, and S. Diaz. 2000. Consequences of changing biodiversity. *Nature* 405(6783):234-242.

Cicin-Sain, B., and R.W. Knecht. 1998. *Integrated Coastal and Ocean Management: Concepts and Practices.* Island Press, Washington, D.C.

Cisneros-Mata, M. A., L. W. Botsford, and J. F. Quinn. 1997. Projecting viability of *Totoaba macdonaldi*, a population with unknown age-dependent variability. *Ecological Applications* 7: 968-980.

Clark, C.W. 1996. Marine reserves and the precautionary management of fisheries. *Ecological Applications* 6:369-370.

Clark, J.R. 1998. *Coastal Seas: The Conservation Challenge.* Blackwell Science, Oxford, U.K.

Clark, J.R., Causey, B., and J.A. Bohnsack. 1989. Benefits from coral reef protection: Looe Key Reef, Florida. Pp. 3076-3086 in O.T. Magoon, H. Converse, D. Minor, L.T. Tobin, and D. Clark (eds.), *Coastal Zone '89. Proceedings of the 6th Symposium on Coastal and Ocean Management, Charleston, S.C., July 11-14.* American Society of Civil Engineers, New York.

Clark, P.A. 1995. Evaluation and management of propeller damage to seagrass beds in Tampa Bay, Florida. *Special Publication of Florida Scientist* 58(2):193-196.

Clark, W.G. 1999. Effects of an erroneous natural mortality rate on a simple age-structured stock assessment. *Canadian Journal of Fisheries and Aquatic Sciences* 10:1721-1731.

Clark, W.G., and S.R. Hare. 1998. Accounting for bycatch in management of the Pacific halibut fishery. *North American Journal of Fisheries Management* 18:809-821.

Cole, R.G., T.M. Ayling, and R.G. Creese. 1980. Effects of marine reserve protection at Goat Island, northern New Zealand. *New Zealand Journal of Marine and Freshwater Research* 24:197-210.

Cole, R.G., T.M. Ayling, and R.G. Creese. 1990. Effects of marine reserve protection at Goat Island, northern New Zealand. *New Zealand Journal of Marine and Freshwater Research* 24 (2): 197-210.

Coleman, F.C., C.C. Koenig, and L.A. Collins. 1996. Reproductive styles of shallow-water groupers (Pisces: Serranidae) in the eastern Gulf of Mexico and the consequences of fishing spawning aggregations. *Environmental Biology of Fishes* 47:129-141.

Coleman, F.C., and J. Travis. In review. Essential fish habitat and marine reserves: A review of the papers emanating from the Second Mote Symposium on Fisheries Ecology. *Ecological Applications.*

Coleman, F.C., C.C. Koenig, A.M. Eklund, and C.B. Grimes. 1999. Management and conservation of temperate reef fishes in the grouper-snapper complex of the southeastern United States. *American Fisheries Society Symposium.* 23: 233-242.

Coleman, F. C., C. C. Koenig, G. R. Huntsman, J. A. Musick, A. M. Eklund, J. C. McGovern, G. R. Sedberry, R. W. Chapman, and C. Grimes. 2000. Long-lived reef fishes: The grouper-snapper complex. *AFS Policy Statement #31c.* American Fisheries Society, Bethesda, MD. (also available on-line at http://www.fisheries.org/Public_Affairs/Policy_Statements/pol_groupersnapper.htm).

Colin, P. L. 1992. Reproduction of the Nassau grouper, *Epinephelus striatus* (Pisces: Serranidae) and its relationship to environmental conditions. *Environmental Biology of Fishes* 34:357-377.

Collie, J.S., G.A. Escanero, P.C. Valentine. 1997. Effects of bottom fishing on the benthic megafauna of Georges bank. *Marine Ecology Progress Series* 155:159-172.

Conover, D.O., J. Travis, and F.C. Coleman. 2000. Essential fish habitat and marine reserves: An introduction to the Second Mote Symposium in Fisheries Ecology. *Bulletin of Marine Science* 66(3):527-534.

Cote, D., D.A. Scruton, G.H. Niezgoda, R.S. McKinley, D.F. Rowsell, R.T. Lindstrom, L.M.N Ollerhead, and C.J. Whitt. 1998. A coded acoustic telemetry system for high precision monitoring of fish location and movement: Application to the study of nearshore nursery habitat of juvenile Atlantic cod (*Gadus morhua*). *Marine Technological Society Journal* 32:54-62.

Cowen, R.K., K.M.M. Lwiza, S. Sponaugle, C.B. Paris, and D.B. Olson. 2000. Connectivity of marine populations: Open or close? *Science* 287:857-859.

Crisp, D.J. 1958. The spread of *Elminius modestus* Darwin in northwest Europe. *Journal of Marine Biology Associations* 37:483-520.

Crowder, L.B., S.J. Lyman, W.F. Figueira, and J. Priddy. 2000. Source-sink population dynamics and the problem of siting marine reserves. *Bulletin of Marine Science* 66(3):799-820.

Cubbage, F.W., J. O'Laughlin, and C.S. Bullock III. 1993. *Forest Resource Policy.* John Wiley & Sons, New York.

Cushing, D.H. 1975. *Marine Ecology and Fisheries.* Cambridge University Press, Cambridge, U.K.

Cushing, D.H. 1988. *The Provident Sea.* Cambridge University Press, Cambridge, U.K.

Daan, N. 1993. Simulation study of effects of closed areas to all fishing, with particular reference to the North Sea ecosystem. Pp. 252-258 in K. Sherman, L.M. Alexander, and B.D. Gold (eds.), *Large Marine Ecosystems: Stress, Mitigation and Sustainability.* American Association for the Advancement of Science Press, Washington, D.C.

Daily, G.C., J.S. Reichert, and J.P. Myers (eds.). 1997. *Nature's Services: Societal Dependence on Natural Ecosystems.* Island Press, Washington, D.C.

Daily, G.C., T. Soderqvist, S. Aniyar, K. Arrow, P. Dasgupta, P.R. Ehrlich, C. Folke, A.M. Jansson, B-O. Jansson, N. Kautsky, S.A. Levin, J. Lubchenco, K-G. Maler, D. Simpson, D. Starrett, D. Tilman, and B. Walker. 2000. The value of nature and the nature of value. *Science* 289:395-396.

Davis, A.R. 1995. Over-exploitation of *Pyura chilensis* (Ascidiacea) in southern Chile—the urgent need to establish marine reserves. *Revista Chilena de Historia Natural* 68:107-116.

Davis, G.E. 1977. Effects of recreational harvest on a spiny lobster, *Panulirus argus*, population. *Bulletin of Marine Science* 27(2):223-236.

Davis, G.E. 1998. Seeking sanctuaries. *National Parks* 72(11-12):41-42

Davis, R. 1963. Recreation planning as an economic problem. *Natural Resources Journal* 3(2):239-249.

Dayton, P.D., and M.J. Tegner, 1990. Bottoms below troubled waters: benthic impacts of the 1982-84 El Niño Southern Oscillation in the temperate zone. Pp. 433-472 in P. Glynn (ed.), *Global consequences of the 1982-83 El Niño Southern Oscillation.* Elsevier Oceanography Series 52. Elsevier, Amsterdam.

Dayton, P.K., S.F. Thrush, M.T. Agardy, and R.J. Hofman. 1995. Environmental effects of marine fishing. *Aquatic Conservation: Marine and Freshwater Ecosystems* 5:205-232.

Dayton, P.K., M.J. Tegner, P.B. Edwards, and K.L. Riser. 1998. Sliding baselines, ghosts, and reduced expectations in kelp forest communities. *Ecological Applications* 8:309-322.

Dayton, P., E. Sala, M.J. Tegner, and S. Thrush. 2000. Marine reserves: Parks, baselines, and fishery enhancement. *Bulletin of Marine Science* 66(3):617-634.

de Forges, B.R., J.A. Koslow, and G.C.B. Poore. 2000. Diversity and endemism of the benthic seamount fauna in the southwest Pacific. *Nature* 405(6789):944-947.

DeMartini, E.E. 1993. Modeling the potential of fishery reserves for managing Pacific coral reef fishes. *Fishery Bulletin* 91:414-427.

DiBacco, C., and L.A. Levin. 2000. Development and application of elemental fingerprinting to track the dispersal of marine invertebrate larvae. *Limnology and Oceanography* 45(4):871-880.

Dixon, J.A., L.F. Scurra, and T. van't Hof. 1993. Meeting ecological and economic goals: Marine parks in the Caribbean. *Ambio* 22(2-3):117-125.

Domeier, M.L., and P.L. Colin. 1997. Tropical reef fish spawning aggregations defined and reviewed. *Bulletin of Marine Science* 60:698-726.

Done, T.J., and R.E. Reichelt. 1998. Integrated coastal zone and fisheries ecosystem management: Generic goals and performance indices. *Ecological Applications* 8(1):S110-S118.

Dufour, V., J.Y. Jouvenel, and R. Galzin. 1995. Study of a Mediterranean reef fish assemblage. Comparisons of population distributions between depths in protected and unprotected areas over one decade. *Aquatic Living Resources* 8(1):17-25.

Dugan, J.E., and G.E. Davis. 1993. Applications of marine refugia to coastal fisheries management. *Canadian Journal of Fisheries and Aquatic Science* 50:2029-2042.

Duke, N.C., M.Z.S. Pinzon, and T.M.C. Prada. 1997. Large-scale damage to mangrove forests following two large oil spills in Panama. *Biotropica* 29:2-14.

Dunlap, R.E., G.H. Gallup, Jr., and A.M. Gallup. 1993a. Of global concern: Results of the health of the planet survey. *Environment* 35(9):7-15, 33.

Dyer, C.L. and J.R. McGoodwin. 1994. *Folk Management in the World's Fisheries: Lessons for Modern Fisheries Management*. University of Colorado Press, Boulder.

Earle, S.A., and W. Henry. 1999. *Wild Ocean: America's Parks Under the Sea*. National Geographic, Washington, D.C.

Ebeling, A.W., and M.A. Hixon. 1991. Tropical and temperate reef fishes: Comparison of community structures. Pp. 509-563 in P.F. Sale (ed.), *The Ecology of Fishes on Coral Reefs*. Academic Press, San Diego, Calif.

Echeverría, J., M. Hanrahan, and R. Solórzano. 1995. Valuation of non-priced amenities provided by the biological resources within the Monteverde Cloud Forest Preserve, Costa Rica. *Ecological Economics* 13:43-52.

Edgar, G.J., and N.S. Barrett. 1997. Short term monitoring of biotic change in Tasmanian marine reserves. *Journal of Experimental Marine Biology and Ecology* 213(2):261-279.

Edwards, S.F. 1991. The demand for Galapagos vacations: Estimation and application to wilderness perservation. *Coastal Management* 19:155-169.

Emanuel, B.P., R.H. Bustamante, G.M. Branch, S. Eekhout, and F.J. Odendaal. 1992. A zoogeographic and functional approach to the selection of marine reserves on the west coast of South Africa. *South African Journal of Marine Science* 12:341-354.

Emlet, R.B., L.R. McEdward, and R.R. Strathman. 1987. Echinoderm larval ecology viewed from the egg. Pp. 55-136 in M. Jangoux and J.M. Lawrence (eds.), *Echinoderm Studies*. A.A. Balkema, Rotterdam, The Netherlands.

Fabrizio, M.C., and R.A. Richards. 1996. Commercial fisheries surveys. Pp. 625-650 in Murphy, B.R. and D.W. Willis (eds.), *Fisheries Techniques*, second edition. American Fisheries Society, Bethesda.

Fagergren, F. 1998. Big Cypress National Preserve: The great compromise. In R. Knight and P. Landres (eds.), *Stewardship Across Boundaries*. Island Press, Covelo, Calif.

Ferguson, L. 1996. *Vessel Traffic Management and the Olympic Coast National Marine Sanctuary*. National Oceanic and Atmospheric Administration, Sanctuaries and Reserves Division, Port Angeles, Wash.

Ferguson, L. 1997. Collaboration for cross-boundary protected area management: Focus on the Olympic Coast National Marine Sanctuary and Olympic National Park. Dissertation, unpublished. University of Washington, Seattle.

Field, J.D. 1997. Atlantic striped bass management: where did we go right? *Fisheries* 22(7): 6-8.

Florida Keys National Marine Sanctuary (FKNMS) 1999. Zone Performance Review. *Second Year Report*. National Oceanic and Atmospheric Administration, Silver Spring, MD.

Florida Sportsman 2000. On the Conservation Front: Block the No Fishing Gang. *Florida Sportsman Magazine* 32 (8).

Fogarty, M.J. 1999. Essential habitat, marine reserves and fishery management. *Trends in Ecology and Evolution* 14(4):133-134.

Fogarty, M.J, and S.A. Murawski. 1998. Large-scale disturbance and the structure of marine ecosystems: Fishery impacts on Georges Bank. *Ecological Applications* 8(1):S63-S71.

Fogarty, M.J., J.A. Bohnsack, and P.K. Dayton. 2000. Marine reserves and resource management. In Sheppard, C (ed.) *Seas at the Millenium*. Elsevier Science Ltd, London.

Food and Agriculture Organization (FAO). 1995. *Precautionary Approach to Fisheries*. FAO Fisheries Technical Report 350. United Nations, Rome.

Food and Agriculture Organization (FAO). 1999. *The State of World Fisheries and Aquaculture 1998*. United Nations, Rome.

Foran, T., and R.M. Fujita. 1999. *Modeling the biological impact of a no-take reserve policy on Pacific continental slope rockfish*. Environmental Defense Fund, Oakland, Calif.

Freeman, A.M. 1993. *The Measurement of Environmental and Resource Values: Theory and Methods*. Resources and Methods, Washington, D.C.

Garcia, S., and C. Newton. 1997. Current situation, trends, and prospects in world capture fisheries. Pp. 3-27 in E.L. Pikitch, D.D. Huppert, and M.P. Sissenwine (eds.), *Global Trends: Fisheries Management*. American Fisheries Society Symposium 20, Bethesda, MD.

Garcia-Rubies, A., and M. Zabala. 1990. Effects of total fishing prohibition on the rocky fish assemblages of Medes Islands Marine Reserve (NW Mediterranean). *Scientia Marina* 54(4):317-328.

Garrod, G., and K. Willis. 1993. The amenity value of a woodland in Great Britain. *Environment and Resource Economics* 2(4):415-434.

Gerber, L.R., S.J. Andelman, L.W. Botsford, S. Gaines, A. Hastings, S. Palumbi and H.P. Possingham. In review. Population models for marine reserve design: a synthesis. *Ecological Applications*.

GESAMP (Joint Group of Experts on the Scientific Aspects of Marine Pollution). 1996. *The Contributions of Science to Integrated Coastal Management*. Report No 61. GESAMP, Rome.

Gibson, C.C., and T. Koontz. 1998. When community is not enough: Communities and forests in southern Indiana. *Human Ecology* 26:621-647.

Gillanders, B.M., and M.J. Kingsford. 1996. Elements in otoliths may elucidate the contribution of estuarine recruitment to sustaining coastal reef populations of a temperate reef fish. *Marine Ecology: Progress Series* 141:13-20.

Gilpin, M., and M.E. Soule. 1986. Minimum Viable Populations: Processes of Species Extinction. Pp. 19-34, in M.E. Soule (ed.), *Conservation Biology: The Science of Scarcity and Diversity*. Sinauer Associates, Sunderland, Massachusetts.

Gittings, S.R., and E.L. Hickerson. 1998. Flower Garden Banks National Marine Sanctuary: Introduction. *Gulf of Mexico Science* 16(2): 128-130.

Glacier Bay National Park (GBNP). 1998. *Glacier Bay National Park, Alaska; Commercial Fishing Regulations and Environmental Assessment*; Proposed Rule, 36 CFR Part 13, National Park Service, Department of the Interior.

Glynn, P.W. 1984. Widespread coral mortality and the 1982/1983 El Niño warming event. *Environmental Conservation* 11:133-146.

Glynn, P.W. 1991. Coral reef bleaching in the 1980s and the possible connections with global warming. *Trends in Ecological Evolution* 6:175-179.

Glynn P.W., and L. D'Croz. 1990. Experimental evidence for high temperature stress as the cause of El Niño-coincident coral mortality. *Coral Reefs* 8:181-191.

Goñi, R. 1998. Ecosystem effects of marine fisheries: An overview. *Ocean Coastal Management* 40:37-64.

Goodridge, R.H., A. Oxenford, B.G. Hatcher, and F. Narcisse. 1997. Changes in the shallow reef fishery associated with implementation of a system of fishing priority and marine reserve areas in Soufriere, St. Lucia. *Proceedings of the 49th Gulf and Caribbean Fisheries Institute* 316-339.

Goodyear, C.P. 1993. Spawning stock biomass per recruit in fisheries management: Foundation and current use. *Canadian Special Publications in Fisheries and Aquatic Science* 120:67-81.

Goodyear, P. 1995. Red snapper in U.S. Waters of the Gulf of Mexico. National Marine Fisheries Service Report Number 95/96-05. Southeast Fisheries Science Center, Miami, Fla.

Goodyear, P. 1997. An evaluation of the minimum reduction in the 1997 red snapper shrimp bycatch mortality rate consistent with the 2019 recovery target. National Marine Fisheries Service, Southeast Fisheries Science Center, Miami, Fla.

Goolsby, D.A. 2000. Mississippi Basin nitrogen flux believed to cause Gulf hypoxia. *EOS* 81(29):321ff.

Grantham, B.A., G.L. Eckert, and A.L. Shanks. In review. Dispersal potential of marine invertebrates in diverse habitats. *Ecological Applications.*

Grassle, J.P., and J.F. Grassle. 1976. Sibling species in the marine pollution indicator *Capitella* (Polychaeta). *Science* 192:567-569.

Grassle, J.F., and J.F. Maciolek. 1992. Deep-sea richness: Regional and local diversity estimates from quantitative bottom samples. *American Naturalist* 139:313-341.

Groombridge, B. (ed.). 1992. *Global Biodiversity: Status of the Earth's Living Resources.* Chapman and Hall, New York.

Guénette, S., and T.J. Pitcher. 1999. An age-structured model showing the benefits of marine reserves in controlling overexploitation. *Fisheries Research* 39:295-303.

Guénette, S., T. Lauck, and C. Clark. 1998. Marine reserves: From Beverton and Holt to the present. *Reviews in Fish Biology and Fisheries* 8:1-21.

Guénette, S., T.J. Pitcher, and C.J. Walters. 2000. The potential of marine reserves for the management of northern cod in Newfoundland. *Bulletin of Marine Science* 66(3):831-852.

Gulland, J.A. 1974. *The Management of Marine Fisheries.* University of Washington Press, Seattle.

Haedrich, R.L. 1995. Structure over time of an exploited deep-water fish assemblage. Pp. 27-50 in A.G. Hopper (ed.), *Deep-Water Fisheries of the North Atlantic.* Kluwer Academic Press, Boston.

Halfpenny, H., and C.M. Roberts. In review. Designing a network of marine reserves for Northwestern Europe. *Ecological Applications.*

Hall, S. 1999. *The Effects of Fishing on Marine Ecosystems and Communities.* Blackwell Science, Malden, Mass.

Halpern, B. In review. The impact of marine reserves: Do reserves work and does reserve size matter? *Ecological Applications.*

Hanley, N., and C. Spash. 1993. *Cost-Benefit Analysis and the Environment.* Edward Elgar, Aldershoot, U.K.

Hanna, S., Blough, H., Allen, R., Iudicello, S., Matlock, G., McCay, B. 2000. *Fishing Grounds: Defining a New Era for American Fisheries Management.* Island Press, Washington, D.C.

Hanna, S. 1998. Institutions for Marine Ecosystems: Economic Incentives and Fishery Management. Ecological Applications 8(1): S170-S174.

Hannesson, R. 1998. Marine reserves: What would they accomplish? *Marine Resource Economics* 13(3):159-170.

Harden Jones, F.R. 1968. *Fish Migration.* Edward Arnold, London.

Hardin, G. 1968. The tragedy of the commons. *Science* 162:1243-1248.

Hardin, G. 1998. Extensions of "the tragedy of the commons." *Science* 280:682-683.

Harmelin, J.G., F. Bachet, and F. Garcia. 1995. Mediterranean marine reserves: Fish indices as tests of protection efficiency. *Marine Ecology* 16:233-250.

Hastings, A., and L. Botsford. 1999. Equivalence in yield from marine reserves and traditional fisheries management. *Science* 284:1-2.

Hawkins, J.P., C.M. Roberts, and V. Clark. 2000. The threatened status of restricted-range coral reef fish species. *Animal Conservation* 3:81-88.

Heyman, W.D., and B. Kjerfve. 1999. Hydrological and oceanographic considerations for integrated coastal zone management in southern Belize. *Environmental Management* 24(2):229-245.

Higgs, A.J., and M.B. Usher. 1980. Should nature reserves be large or small? *Nature* 285: 568-569.

Hilborn, R., and C.J. Walters. 1992. *Quantitative Fisheries Stock Assessment: Choice Dynamics and Uncertainty.* Routledge, Chapman & Hall, New York.

Hilborn, R., J.J. Maguire, A.M. Parma, and A.A. Rosenberg. In press. The precautionary approach and risk management. Can they increase the probability of success in fishery management? *Special Publication of the Canadian Journal of Fisheries and Aquatic Science.*

Hjort, J. 1914. Fluctuations in the great fisheries of northern Europe. Rapports et Proces. *Verbaux du Conseil International pour L'Exploration de la Mer* 20:1-228.

Hoagland, P., Y. Kaoru and J.M. Broadus. 1995. A methodological review of net benefit evaluation for marine reserves. Environmental Department Paper no. 27, The World Bank, Washington, D.C.

Hockey, P.A., and G.M. Branch. 1994. Conserving marine biodiversity on the African coast: Implications of a terrestrial perspective. *Aquatic Conservation: Marine Freshwater Ecosystems* 4(4):345-362.

Hockey, P.A., and G.M. Branch. 1997. Criteria, objectives and methodology for evaluating marine protected areas in South Africa. *South African Journal of Marine Science* 18:369-383.

Holland, D.S., and R.J. Brazee. 1996. Marine reserves for fisheries management. *Marine Resource Economics* 11:157-171.

Honneland, G. 1999. Co-management and communities in the Barents Sea fisheries. *Human Organization* 58(4):397-404.

Horwood, J.W., J.H. Nichols, and S. Milligan. 1998. Evaluation of closed areas for fish stock conservation. *Journal of Applied Ecology* 35:893-903.

Houde, E.D. 1987. Fish early life dynamics and recruitment variability. *American Fisheries Society Symposium* 2:17-29.

Houde, E. D. 1997. Patterns and consequences of selective processes in teleost early life histories. Pp. 173-196. In: Chambers, R. C. and E. A. Trippel (eds.), *Early Life History and Recruitment in Fish Populations.* Chapman & Hall, London

Hughes, T.P. 1994. Catastrophes, phase shifts, and large-scale degradation of a Caribbean coral reef. *Science* 265:1547-1551.

Hundloe, T. 1989. Measuring the value of the Great Barrier reef. *Australian Parks and Recreation* 3(26):11-15.

Hutchings, J.A. 2000. Collapses and recovery of marine fishes. *Nature* 406:882-885.

Inglehart, R. 1990. *Culture Shift in Advanced Industrial Society.* Princeton University Press, Princeton, N.J.

Inglehart, R. 1991. *Eurobarometer: The Dynamics of European Public Opinion: Essays in Honour of Jacques-René Rabier.* Macmillan, London.

Inglehart, R. 1997. *Modernization and Postmodernization: Cultural, Economic, and Political Change in 43 Societies.* Princeton University Press, Princeton, N.J.

International Union for the Conservation of Nature and Natural Resources (IUCN). 1976. *An International Conference on Marine Parks and Reserves: Papers and Proceedings of an International Conference held in Tokyo, Japan, May 12-14 1975.* IUCN, Morges, Switzerland.

International Union for the Conservation of Nature and Natural Resources (IUCN). 1980. *World Conservation Strategy: Living Resource Conservation for Sustainable Development.* IUCN, Gland, Switzerland.

International Union for the Conservation of Nature and Natural Resources (IUCN). 1987. P.R. Dingwall (ed.), *Proceedings of the 29th Working Session of IUCN's Commission on National Parks and Protected Areas, Wairakei, New Zealand, August 16-21, 1987.* IUCN, Gland, Switzerland.

International Union for the Conservation of Nature and Natural Resources (IUCN). 1994. *Guidelines for Protected Area Management Categories.* IUCN, Gland, Switzerland, and Cambridge, U.K.

Iverson, E.S. 1996. *Living Marine Resources.* Chapman and Hall, New York.

Jackson, G.A., and R.R. Strathmann. 1981. Larval mortality from offshore mixing as a link between precompetent and competent periods of development. *American Naturalist* 118:16-26.

Jameson, S.C., M.V. Erdmann, G.R. Gibson, Jr., and K.W. Potts. 1998. Development of biology criteria for coral reef ecosystem assessment. *Atoll Research Bulletin* 450 (September).

Jennings, S., and N.V.C. Polunin. 1997. Impacts of predator depletion by fishing on the biomass and diversity of non-target reef fish communities. *Coral Reefs* 16:71-82.

Jennings, S., E.M. Grandcourt, and N.V.C. Polunin. 1995. The effects of fishing on the diversity, biomass, and trophic structure of Seychelles' reef fish communities. *Coral Reefs* 14:225-235.

Jennings, S., S.S. Marshall, and N.V.C. Polunin. 1996. Seychelles' marine protected areas: Comparative structure and status of reef fish communities. *Biological Conservation* 75(3):201-209.

Johannes. R.E. 1978. Traditional marine conservation methods in oceania and their demise. *Annual Reviews of Ecology and Systematics* 9:349-64.

Johannes, R.E. 1998. The case for data-less marine resource management: Examples from tropical nearshore finfisheries. *Trends in Ecology and Evolution* 13:243-246.

Johnson, D., N. Funicelli, and J. Bohnsack. 1999. Effectiveness of an existing estuarine no-take fish sanctuary within the Kennedy Space Center, Florida. *North American Journal of Fisheries Management* 19:436-453.

Johnson, M.S., and R. Black. 1998. Effects of isolation by distance and geographical discontinuity on genetic subdivison of *Littorina cingulata. Marine Biology* 132:295-303.

Jones, G.P., M.J. Milicich, M.J. Emslie, and C. Lunow. 1999. Self-recruitment in a coral reef fish population. *Nature* 402:802-804.

Jones, J.B. 1992. Environmental impact of trawling on the seabed: a review. *New Zealand Journal of Marine and Freshwater Research* 26:59-67.

Kaiser, M.J. 1998. Significance of bottom-fishing disturbances. *Conservation Biology* 12:1230-1235.

Kaoru, Y. 1993. Differentiating use and nonuse values for coastal pond water quality improvements. *Environmental and Resource Economics* 3:487-494.

Kaoru, Y., and J.M. Broadus. 1994. A socioeconomic evaluation of marine resource uses in Wellfleet Harbor. Mimeo. Woods Hole: Woods Hole Research Consortium.

Keiter, R. 1988. Ecosystem management for parks and wilderness. Pp. 15-20 in J. Agee and D. Johnson (eds.), *Natural Ecosystem Management in Park and Wilderness Areas: Looking at the Law.* University of Washington Press, Seattle.

Kelleher, G. 1999. *Guidelines for Marine Protected Areas.* International Union for the Conservation of Nature and Natural Resources, Gland, Switzerland and Cambridge, U.K.

Kelleher, G., and R. Kenchington. 1982. Australia's Great Barrier Reef Marine Park: Making development compatible with conservation. *Ambio* 11:262-267.

Kelleher, G., and R. Kenchington. 1992. *Guidelines for Establishing Marine Protected Areas. A Marine Conservation and Development Report.* World Conservation Union (IUCN), Gland, Switzerland.

Kelleher, G., C. Bleakley, and S. Wells. 1995. *A Global Representative System of Marine Protected Areas.* The Great Barrier Reef Marine Park Authority, the World Bank, and International Union for the Conservation of Nature and Natural Resources, Washington, D.C.

Kelleher, G., and C. Recchia. 1998. Editorial – Lessons from marine protected areas around the world. *Parks* 8: 1-4.

Kempton, W., J.S. Boster, and J.A. Hartley. 1995. *Environmental Values in American Culture.* Massachusetts Institute of Technology Press, Cambridge.

Kenchington, E., M. Willison, M. Butler, J. Hall, P. Doherty, S. Fuller, and S. Gass. In press. *Proceedings of the First International Symposium on Deep Sea Corals,* Halifax, Nova Scotia, August 2000. Ecology Action Centre and Nova Scotian Institute of Science Special Publication. Halifax, Nova Scotia.

Keough, M.J., G.P. Quinn, and A. King. 1993. Correlations between human collecting and intertidal mollusc populations on rocky shores. *Conservation Biology* 7:378-390.

Klee, G.A. 1999. *The Coastal Environment: Toward Integrated Coastal and Marine Sanctuary Management.* Prentice Hall, Upper Saddle River, NJ.

Knowlton, N. 1993. Sibling species in the sea. *Annual Review of Ecology and Systematics* 24: 189-216.

Koenig, C.C., and F.C. Coleman. 1998. Absolute abundance and survival of juvenile gags in sea grass beds of the northeastern Gulf of Mexico. *Transactions of the American Fisheries Society* 127:44-55.

Koenig, C.C., F.C. Coleman, C.B. Grimes, G. Fitzhugh, C. Gledhill, K.M. Scanlon, and M. Grace. 2000. Protection of fish spawning habitat for conservation of warm-temperate reef fish fisheries of shelf-edge reefs of Florida. *Bulletin of Marine Science* 66(3):593-616.

Koslow, J.A. 1997. Seamounts and the ecology of deep-sea fisheries. *American Scientist* 85:168-176.

Kramer, D.L., and M.R. Chapman. 1999. Implications of fish home range size and relocation for marine reserve function. *Environmental Biology of Fishes* 55:65-79.

Kronman, M. 1999. No parking. *National Fisherman* 80(2):22-24.

Krutilla, J. 1967. Conservation reconsidered. *American Economic Review* 57:787-796.

Lafferty, K.D., J.E. Dugan, H. Leslie, D. McArdle, and R. Warner. In review. Applying integrative marine reserve design: Examples of options for the California Channel Islands. *Ecological Applications*

Langton, R.W., and P.J. Auster. 1999. Marine fishery and habitat interactions: To what extent are fisheries habitat interdependent? *Fisheries* 24(6):14-21.

Larkin, P.A. 1977. An epitaph for the concept of MSY maximum sustainable yield. *Transactions of the American Fisheries Society* 106:1-11.

Lasker, R. 1975. Field criteria for survival of anchovy larvae: The relation between inshore chlorophyll maximum layers and successful first feeding. *Fisheries Bulletin* 73:453-462.

Lauck, T. 1996. Uncertainty in fisheries management. Pp. 91-105 in D.V. Gordon and G.R. Munro (eds.), *Fisheries and Uncertainty: A Precautionary Approach to Resource Management.* University of Calgary Press, Canada.

Lauck, T.C., C.W. Clark, M. Mangel, and G.R. Munro. 1998. Implementing the precautionary principle in fisheries management through marine reserves. *Ecological Applications* 8(1):S72-S78.

Lee, T.N., M.E. Clarke, E. Williams, A.F. Szmant, and P. Berger. 1994. Evolution of the Tortugas gyre and its influence on recruitment in the Florida Keys. *Bulletin of Marine Science* 54(3):625-646.

Leeworthy, V.R. 1991. The feasibility of user fees in national marine sanctuaries: A preliminary characterization. Mimeo. Washington: Strategic Environmental Assessments Division, National Oceanic and Atmospheric Administration (December).

Leslie, H., M. Ruckleshaus, I. Ball, S. Andelman, and H. Possingham. In review. Using siting algorithms in the design of marine reserves. *Ecological Applications.*

Letourneur, Y. 1996. Responses of fish populations to marine reserves: The case of Mayotte Island, western Indian Ocean. *Ecoscience* 3(4):442-450.

Levin, L.A., D. Huggett, P. Myers, T. Bridges and J. Weaver. 1993. Rare-earth tagging methods for the study of larval dispersal by marine invertebrates. *Limnology and Oceanography* 38:346-360.

Levitan, D.R. 1998. Sperm limitation, gamete competition, and sexual selection in external fertilizers. In *Sperm Competition and Sexual Selection*. Academic Press, Washington, D.C.

Levitan, D.R., M.A. Sewell, F.S. Chia. 1992. How distribution and abundance influences fertilization success in the sea urchin *Strongylocentrotus franciscanus*. *Ecology* 73:248-254.

Lillie, F.R. 1915. Studies of fertilization. VII. Analysis of variations in the fertilization power of sperm suspension of *Arbacia*. *Biological Bulletin* 28:229-251.

Limouzy-Paris, C.B., H.C. Graber, D.L. Jones, A.W. Ropke, and W.J. Richards. 1997. Translocation of larval coral reef fishes via sub-mesoscale spin-off eddies from the Florida Current. *Bulletin of Marine Science* 60:966-983.

Lindeman, K.C., and D.B. Snyder. 1999. Nearshore hardbottom fishes of southeast Florida and effects of habitat burial caused by dredging. *Fisheries Bulletin* 97:508-525.

Lindholm, J. B., P. J. Auster, and L. S. Kaufman. 1999. Habitat-mediated survivorship of juvenile (0-year) Atlantic cod *Gadus morhua*. *Marine Ecology Progress Series* 180:247-255.

Lipcius, R.N., W.T. Stockhausen, D.B. Eggleston, L.J. Marshall, and B. Hickey. 1997. Hydrodynamic decoupling of recruitment, habitat quality, and adult abundance in the Caribbean spiny lobster: Source-sink dynamics. *Marine Freshwater Research* 49:317-323.

Longhurst, A. 1998. *Ecological Geography of the Sea*. Academic Press, London.

Ludwig, D., R. Hilborn, and C. Walters. 1993. Uncertainty, resource exploitation, and conservation. *Science* 260:17,36.

MacDiarmid, A.B., and P.A. Breen. 1992. Spiny lobster population changes in a marine reserve. Pp. 47-56 in C.N. Battershill et al. (eds.), *Proceedings of the International Temporal Reef Symposium*. NIWA Marine, Wellington, New Zealand.

Mace, P.M. 1994. Relationships between common biological reference points used as thresholds and targets of fisheries management strategies. *Canadian Journal of Fisheries and Aquatic Science* 51:110-122.

Mace, P.M., and M.P. Sissenwine. 1993. How much spawning per recruit is enough? *Canadian Special Publication of Fisheries and Aquatic Sciences* 120:101-118.

Maloney, C.L., L.W. Botsford, and J.L. Largier. 1994. Development, survival and timing of metamorphosis of planktonic larvae in a variable environment: The Dungeness crab as an example. *Marine Ecology Progress Series* 113:61-79.

Man, A., R. Law, and N.V.C. Polunin. 1995. Role of marine reserves in recruitment to reef fisheries: A metapopulation model. *Biological Conservation* 71:197-204.

Mangel, M. 2000. Trade-offs between fish habitat and fishing mortality and the role of reserves. *Bulletin of Marine Science* 66(3):663-674.

Mantell, M. A., and P. C. Metzgar. 1990. The Organic Act and the stewardship of resources within park boundaries. Chapter 2 in M.A. Mantell (ed.), *Managing the National Park System: A Handbook of Legal Duties, Opportunities, and Tools*. The Conservation Foundation, Washington, D.C.

Marine Fish Conservation Network and Center for Marine Conservation. 1999. *Missing the Boat: An Evaluation of Fishery Management Response to the Sustainable Fisheries Act*. Marine Fish Conservation Network and the Center for Marine Conservation, Washington, D.C.

Marine Protected Area News (MPA News). 1999. California passes law to network its MPAs, create no-take reserves. *MPA News* 1(3):1.

Marsh, A.G., and D.T. Manahan. 1999. A method for accurate measurements of the respiration rates of marine invertebrate embryos and larvae. *Marine Ecology Progress Series* 184:1-10.

Marsh, A.G., P.K.K. Leong, and D.T. Manahan. 1999. Energy metabolism during embryonic development and larval growth of an Antarctic sea urchin. *Journal of Experimental Biology* 202:2041-2050.

McAllister, D.E., J. Baquero, G. Spiller, and R. Campbell. 1999. *A Global Trawling Ground Survey*. Marine Conservation Biology Institute, Seattle, World Resources Institute, Washington D.C, and Ocean Voice International, Ottawa.

McAllister, M. 1997. Peer review of red snapper (*Lutjanus campechanus*) research and management in the Gulf of Mexico. Unpublished paper, submitted to the National Oceanic and Atmospheric Administration's National Marine Fisheries Service.

McArdle, D. 1997. *California Marine Protected Areas*. Publication Number T-039. California Sea Grant College System, La Jolla, Calif.

McCay, B. 2000. Sea changes in fisheries policy: Contributions from anthropology. Pp. 201-217 in E.P. Durrenberger and T.P. King (eds.), *State and Community in Fisheries Management: Power, Policy, and Practice*. Bergin & Garvey, Westport, Conn.

McCay, B., and J. Acheson (eds.). 1987. *The Question of the Commons: The Culture and Ecology of Communal Resources*. University of Arizona Press, Tucson.

McClanahan, T.R. 1989. Kenyan coral reef-associated gastropod fauna: A comparison between protected and unprotected reefs. *Marine Ecology Progress Series* 53:11-20.

McClanahan, T.R. 1990. Viewpoint: Are conservationists fish bigots? *BioScience* 40(1):2.

McClanahan, T.R. 1995. A coral reef ecosystem-fisheries model: Impacts of fishing intensity and catch selection on reef structure and processes. *Ecological Modeling* 80:1-19.

McClanahan, T.R. 1999. Is there a future for coral reef parks in poor tropical countries? *Coral Reefs* 18(4):321-325.

McClanahan, T.R., and B. Kaunda-Arara. 1996. Fishery recovery in a coral-reef marine park and its effect on the adjacent fishery. *Conservation Biology* 10(4):1187-1199.

McClanahan, T.R., M. Nugues, and S. Mwachireya. 1994. Fish and sea urchin herbivory and competition in Kenyan coral reef lagoons: The role of reef management. *Journal of Experimental Marine Biology and Ecology* 184:237-254.

McClanahan, T.R., and S.H. Shafir. 1990. Causes and consequences of sea urchin abundance and diversity in Kenyan coral reef lagoons. *Oecologia* 362-370.

McCormick, M.I., and J.H. Choat. 1987. Estimating total abundance of a large temperate reef fish using visual strip transects. *Marine Biology* 96:469-478.

McDonnell, M.J., and S.T.A. Pickett (eds.). 1993. *Humans as Components of Ecosystems*. Springer-Verlag, New York.

McGarvey, R., and J.H.M. Willison. 1995. Rationale for a marine protected area along the international boundary between U.S. and Canadian waters in the Gulf of Maine. Pp. 74-81 in N.L. Shackell and J.H.M. Willison (eds.), *Marine Protected Areas and Sustainable Fisheries*. Science and Management of Protected Areas Association, Wolfville, Canada.

McGovern, J.C., D.M. Wyanksi, O. Pashuk, C.S. Manooch III, and G.R. Sedberry. 1998. Changes in the sex ratio and size at maturity of gag, *Mycteroperca microlepis*, from the Atlantic Coast of the southeastern United States during 1976-1995. *Fisheries Bulletin* 96:797-807.

McManus, J.W., and L.A.B. Meñez. 1997. The proposed international Spratly Island marine park: ecological considerations. *Proceedings from the 8th International Coral Reef Symposium, Panama* 2:1943-1948.

McShane P.E., and M.G. Smith. 1988. Measuring abundance of juvenile abalone, *Haliotis rubra*, Leach (gastropoda, haliotidae): Comparison of a novel method with two other methods. *Australian Journal of Marine Freshwater Research* 39(3):331-336.

McShane, P.E., K.P. Black, and M.G. Smith. 1988. Recruitment processes in *Haliotis rubra* (Mollusca: Gastropoda) and regional hydrodynamics in southeastern Australia imply localized dispersal of larvae. *Journal of Experimental Marine Biology and Ecology* 124 (3): 175-203.

Merret, N.R., and Haedrich, R.L. 1997. *Deep Sea Fish and Fisheries*. Chapman and Hall, London.

Merton, Robert K. 1957. *Social Theory and Social Structure*. Free Press, New York.

Metcalfe, J.D., and G.P. Arnold. 1997. Tracking fish with electronic tags. *Nature* 387:665-666.

Methot, R.D. 1986. Management of Dungeness crab fisheries. In G.S. Jamieson and N. Bourne (eds.), *North Pacific Workshop on Stock Assessment and Management of Invertebrates. Canadian Special Publication of Fisheries and Aquatic Science* 92:326-334.

Mills, M. 1999. *Strategy and Recommended Action List for Protection and Restoration of Marine Life in the Inland Waters of Washington State.* Puget Sound-Georgia Basin International Task Force, Olympia, Wash.

Milon, J. W. 2000. Pastures, fences, tragedies and marine reserves. *Bulletin of Marine Science* 66(3):901-916.

Minns, C.K., R.G. Randall, J.E. Moore, and V.W. Cairns. 1996. A model simulating the impact of habitat supply limits on northern pike, Essox lucius, in Hamilton Harbour, Lake Ontario. *Canadian Journal of Fisheries and Aquatic Sciences* 53: 20-34.

Mitchell, R., and R. Carson. 1989. *Using Surveys to Value Public Goods: The Contingent Value Method.* Resources for the Future, Washington, D.C.

Moore, J.A. 1999. Deep-sea finfish fisheries: Lessons from history. *Fisheries* 24:16-21.

Moribe, J.T. 1999. Visitor attitudes toward the prohibition of fish feeding in the Hanauma Bay Marine Life Conservation District. Dissertation, unpublished. University of Washington, Seattle.

Morton, B. 1996. The subsidiary impacts of dredging (and trawling) on a subtidal benthic molluscan community in the southern waters of Hong Kong. *Marine Pollution Bulletin* 32(10):701-710.

Murawski, S.A., R. Brown, H.-L. Lai, P.J. Rago, and L. Hendrickson. 2000. Large-scale closed areas as a fishery-management tool in temperate marine systems: The Georges Bank experience. *Bulletin of Marine Science* 66(3):775-798.

Murray, M., and L. Ferguson. 1998. *The Status of Marine Protected Areas in Puget Sound.* Volumes I and II. Puget Sound-Georgia Basin Environmental Report Series: Number 8. Seattle, Wash.

Murray, P, and J. Metcalf. 1998. *Murray-Metcalf Northwest Straits Citizens Advisory Commissions: Report to the Convenors.* Washington Sea Grant, Seattle.

Murray, S.N., R.F. Ambrose, J.A. Bohnsack, L.W. Botsford, M.H. Carr, G.E. Davis, P.K. Dayton, D. Gotshall, D.R. Gunderson, M.A. Hixon, J. Lubchenco, M. Mangel, A. MacCall, D.A. McArdle, J.C. Ogden, J. Roughgarden, R.M. Starr, M.J. Tegner, and M.M. Yoklavich. 1999. No-take reserve networks: Sustaining fishery populations and marine ecosystems. *Fisheries* 24(11):11-24.

Musick, J.A. 1999. Ecology and conservation of long-lived marine animals. In J. A. Musick (ed.) Life in the slow lane: Ecology and conservation of long-lived marine animals. *American Fisheries Society Symposium* 23:1-10.

Myers, R.A., J.A. Hutchings, and N.J. Barrowman. 1996. Hypotheses for the decline of cod in the North Atlantic. *Marine Ecology Progress Series* 138:293-308.

Myers, R.A., J.A. Hutchings, and N.J. Barrowman. 1997. Why do fish stocks collapse? The examples of cod in Atlantic Canada. *Ecological Applications* 7(1):91-106.

Myers, R.A., and G. Mertz. 1998. The limits of exploitation: a precautionary approach. *Ecological Applications.* 8(1):S165-S169.

Nagelkerken, W.P. 1981. Distribution and ecology of the groupers (Serranidae) and snappers (Lutjanidae) of the Netherlands Antilles. *Natuurwetenschappelijke Studiekring voor Suriname en de Nederlandse Antillen* 107:1-71.

National Academy of Public Administration (NAPA). 2000. *Protecting Our National Marine Sanctuaries.* National Academy of Public Administration, Washington, D.C.

National Marine Fisheries Service (NMFS). 1999. *Ecosystem-based Fishery Management.* Ecosystem Advisory Panel to NMFS. NOAA Technical Memorandum NMFS-F/SPO-33. National Marine Fisheries Service, Silver Spring, MD.

National Oceanic and Atmospheric Administration (NOAA). 1996a. *Magnuson-Stevens Fishery Conservation and Management Act.* NOAA Technical Memorandum NMFS-F/SPO-23. National Oceanic and Atmospheric Administration, Silver Spring, MD.

National Oceanic and Atmospheric Administration (NOAA). 1996b. *Our Living Oceans: Report on the Status of U.S. Living Marine Resources, 1995.* NOAA Technical Memorandum NMFS-F/SPO-19. National Oceanic and Atmospheric Administration, Silver Spring, MD.

National Oceanic and Atmospheric Administration (NOAA). 1997. *Accomplishments Report.* National Marine Sanctuary Program, Silver Spring, MD.

National Oceanic and Atmospheric Administration (NOAA). 1999. *Our Living Oceans: Report on the Status of U.S. Living Marine Resources, 1999.* NOAA Technical Memorandum NMFS-F/SPO-41. National Oceanic and Atmospheric Administration, Silver Spring, MD.

National Oceanic and Atmospheric Administration, Sanctuaries and Reserves Division (NOAA SRD). 1996a. *1996 Program Overview for the National Marine Sanctuaries and National Estuarine Research Reserves.* National Oceanic and Atmospheric Administration, Silver Spring, MD.

National Oceanic and Atmospheric Administration, Sanctuaries and Reserves Division (NOAA SRD). 1996b. *Florida Keys National Marine Sanctuary: Final management plan/environmental impact statement, vol. 1.* National Oceanic and Atmospheric Administration, Silver Spring, MD.

National Park Service (NPS). 1998. *Coral Reefs Under National Park Service Jurisdiction: Overview of Areas, Protection, and Management Issues.* U.S. Department of the Interior, Water Resources Division, Washington, D.C.

National Research Council (NRC). 1990. *Managing Troubled Waters: The Role of Marine Environmental Monitoring.* National Academy Press, Washington, D.C.

National Research Council (NRC). 1992. *Restoration of Aquatic Ecosystems: Science, Technology, and Public Policy.* National Academy Press, Washington, D.C.

National Research Council (NRC). 1995. *Understanding Marine Biodiversity.* National Academy Press, Washington, D.C.

National Research Council (NRC). 1997. *Striking A Balance: Improving Stewardship of Marine Areas.* National Academy Press, Washington, D.C.

National Research Council (NRC). 1998a. *Improving Fish Stock Assessments.* National Academy Press, Washington, D.C.

National Research Council (NRC). 1998b. *Review of Northeast Fishery Stock Assessments.* National Academy Press, Washington, D.C.

National Research Council (NRC). 1999a. *Sustaining Marine Fisheries.* National Academy Press, Washington, D.C.

National Research Council (NRC). 1999b. *Sharing the Fish.* National Academy Press, Washington, D.C.

National Research Council (NRC). 2000a. *Bridging Boundaries Through Regional Marine Research.* National Academy Press, Washington, D.C.

National Research Council (NRC). 2000b. *Clean Coastal Waters: Understanding and Reducing the Effects of Nutrient Pollution.* National Academy Press, Washington, D.C.

National Research Council (NRC). 2000c. *Improving the Collection, Management, and Use of Marine Fisheries Data.* National Academy Press, Washington, D.C.

Naylor, R.L., R.J. Goldburg, H. Mooney, M. Beveridge, F. Clay, C. Folke, N. Kautsky, J. Lubchenco, J. Primavera, and M. Williams. 1998. Nature's subsidies to shrimp and salmon farming. *Science* 282:883-884.

Naylor, R.L., R.J. Goldburg, J. Primavera, N. Kautsky, M. Beveridge, J. Clay, C. Folke, J. Lubchenco, H. Mooney, and M. Troell. 2000. Effect of aquaculture on world fish supplies. *Nature* 405:1017-1024.

New, M.B. 1997. Aquaculture and the capture fisheries: balancing the scales. *World Aquaculture* 28:11.

Nichols, S., and G.J. Pellegrin. 1992. Revision and update of estimates of shrimp fleet bycatch, 1972-1991. Report to the Gulf of Mexico Fishery Management Council, National Marine Fisheries Service, Pascagoula, Fla.

Nichols, S., A. Shah, G.J. Pellegrin, Jr., and K. Mullin. 1990. Updated estimates of shrimp fleet bycatch in the offshore waters of the U.S. Gulf of Mexico 1972-1989. Unpublished report. U.S. Department of Commerce, NOAA, NMFS Southeast Fisheries Science Center, Pascagoula, Fla.

Nilsson, P. 1998. *Criteria for the Selection of Marine Protected Areas.* Report 4834. Swedish Environmental Protection Agency. Stockholm.

Nordiska Ministerrådet. 1995. Marina reservat i Norden. Del I (Marine protected areas in the Nordic countries. Part I). Nordiska Ministerrådet (Nordic Council of Ministers) Köpenhamn. *TemaNord* 1995:553.

Ogden, J.C. 1997. Marine managers look upstream for connections. *Science* 278:1414-1415.

Olson, M. 1965. *The Logic of Collective Action.* Harvard University Press, Cambridge, Mass.

Olson. 1985. The consequences of short-distance larval dispersal in a sessile marine invertebrate. *Ecology* 66:30-39.

Oregon Ocean Policy Advisory Council (OOPAC). 1994. *State of Oregon: Territorial Sea Plan.* Oregon Ocean Policy Advisory Council, Portland.

Orensanz, J.M., and G.S. Jamieson. 1998. The assessment and management of spatially structured stocks: An overview of the north Pacific symposium on invertebrate stock assessment and management. Pp. 441-45 in G.S. Jamieson and A. Campbell (eds.), *Proceedings of the North Pacific Symposium on Invertebrate Stock. Canadian Special Publication of Fisheries and Aquatic Sciences* 125:441-459.

Ostrom, E. 1990. *Governing the Commons: the Evolution of Institutions for Collective Action.* Cambridge University Press, Cambridge, U.K.

Paddack, M.J. 1996. The influence of marine reserves upon rockfish populations in central California kelp forests. M.S. thesis. University of California, Santa Cruz.

Palumbi, S.R. 1992. Marine speciation on a small planet. *Trends in Ecology and Evolution* 7:114-118.

Palumbi, S.R. 1994. Reproductive isolation, genetic divergence, and speciation in the sea. *Annual Review of Ecology and Systematics* 25:547-572.

Palumbi, S.R. 1997. Molecular biogeography of the Pacific. *Coral Reefs* 16:S47-S52.

Palumbi, S.R. 2000. The ecology of marine protected areas. In M. Bertness, M. Hixon, and S. Gaines (eds.), *Marine Community Ecology.* Sinauer Press, Sunderlan, Mass.

Palumbi, S.R. In review. Population genetics, demographic connectivity and the design of marine reserves. *Ecological Applications.*

Parma, A.M. 1993. Retrospective catch-at-age analysis of Pacific halibut: Implications on assessment of harvesting policies. Pp. 247-265 in G. Kruse, D.M. Eggers, R.J. Marasco, C. Pautzke, and T.J. Quinn II (eds.), *Proceedings of the International Symposium on Management Strategies for Exploited Fish Populations.* Alaska Sea Grant College Program Report No. 93-02. University of Alaska, Fairbanks.

Parma, A.M., and R.B. Deriso. 1990. Dynamics of age and size in a population subject to size-selective mortality: effects of phenotypic variability in growth. *Canadian Journal of Fisheries and Aquatic Sciences* 47:274-289.

Perry, R.I., C.J. Walters, and J.A. Boutillier. 1999. A framework for providing scientific advice for the management of new and developing invertebrate fisheries. *Reviews in Fish Biology and Fisheries* 9:125-150.

Pezzey, J.C.V., C.M. Roberts, and B.T. Urdal. In press. A simple bioeconomic model of a marine reserve. *Ecological Economics.*

Pilskaln, C.H., J.H. Churchill, and L.M. Mayer. 1998. Frequency of bottom trawling in the Gulf of Maine and speculations on the geochemical consequences. *Conservation Biology* 12:1223-1224.

Pimm, S.L. 1982. *Food Webs.* Chapman and Hall, London.

Pinkerton, E.W. 1989. *Co-operative Management of Local Fisheries: New Directions for Improving Management and Community Development.* University of British Columbia Press, Vancouver.

Pinkerton, E.W. 1994. Local fisheries co-management: A review of international experiences and their implications for salmon management in British Columbia. *Canadian Journal of Fisheries and Aquatic Sciences* 51:1-17.

Poizat, G., and E. Baran. 1997. Fishermen's knowledge as background information in tropical fish ecology: A quantitative comparison with fish sampling results. *Environmental Biology of Fishes* 50:435-449.

Polacheck, T. 1990. Year around closed areas as a management tool. *Natural Resource Modeling* 4:327-354.

Policansky, D. 1993. Evolution and management of exploited fish populations. In G. Kruse, D.M. Eggers, R.J. Marasco, C. Pautzke, and T.J. Quinn II (eds.), *Proceedings of the International Symposium on Management Strategies for Exploited Fish Populations.* Alaska Sea Grant College Program Report No. 93-02. University of Alaska, Fairbanks.

Polunin, N.V.C., and C.M. Roberts. 1993. Greater biomass and value of target coral reef fishes in two small Caribbean marine reserves. *Marine Ecology Progress Series* 100:167-176.

Pope, J.A. 1988. Collecting fisheries assessment data. Pp. 63-82 in Gulland, J.A. (ed.), *Fish Population Dynamics*, second edition. J. Wiley & Sons, Chichester.

Poss, S.G. 1998. Species at risk in the Gulf of Mexico ecosystem, a Web-site (http://lionfish.ims.usm.edu/~musweb/endanger.html). A cooperative program between the US Gulf of Mexico Program and the Gulf Coast Research Laboratory Museum to identify potentially endangered species in the Gulf of Mexico and to determine research needs for these species. Hattiesburg, MS.

Pressey, R.L, C.J. Humphries, C.R. Margules, R.I. Vane-Wright, and P.H. Williams. 1993. Beyond opportunism: key principles for systematic reserve selection. *Trends in Ecology and Evolution* 8: 124-128.

Pulliam, H.R. 1988. Sources, sinks, and population regulation. *American Naturalist* 132:652-661.

Quinn, J., S.R. Wing, and L.W. Botsford. 1993. Harvest refugia in marine invertebrate fisheries: Models and applications to the red sea urchin, *Strongylocentrotus franciscanus*. *American Zoologist* 33:537-550.

Rakitin, A., and D.L. Kramer. 1996. Effect of a marine reserve on the distribution of coral reef fishes in Barbados. *Marine Ecology Progress Series* 131(1-3):97-113.

Ray, G.C. 1999. Coastal-marine protected areas: Agonies of choice. *Aquatic Conservation: Marine and Freshwater Ecosystems* 9:607-614.

Reef Fisheries Plan Development Team (RFPDT). 1990. *The Potential of Marine Fishery Reserves for Reef Management in the U.S. South Atlantic.* NOAA Technical Memorandum NMFS-SEFC-261. Contribution No. CRD/89-90/04. National Oceanic and Atmospheric Administration, Silver Spring, MD.

Reinert, T.R., J. Wallin, M.C. Conroy, M.J. Avyle, and V.D. Avyle. 1998. Long-term retention and detection of oxytetracycline marks applied to hatchery-reared larval striped bass (*Morone saxatilis*). *Canadian Journal of Fisheries and Aquatic Science* 55:539-543.

Restrepo, V.R., and J.E. Powers. 1999. Precautionary control rules in United States fisheries management: Specification and performance. *ICES Journal of Marine Science* 56:846-852.

Rice, M.A., C. Hickox, and I. Zehra. 1989. Effects of intensive fishing effort on the population structure of quahogs, *Mercenaria mercenaria* (Linnaeus 1758), in Narragansett Bay. *Journal of Shellfish Research* 8:345-354.

Richards, R.A., and P.J. Rago. 1999. A case history of effective fishery management: Chesapeake Bay striped bass. *North American Journal of Fisheries Management* 19:356-375.

Ricker, W.E. 1981. Changes in the average size and average age of Pacific salmon. *Canadian Journal of Fisheries and Aquatic Science* 38: 1636-1656.

Ridker, R.G., and J.A. Henning. 1967. The determinants of residential property values with special reference to air pollution. *Review of Economics and Statistics* 49:246-257.

Roberts, C.M. 1991. Larval mortality and the composition of coral reef fish communities. *Trends in Ecology and Evolution* 6:83-87.

Roberts, C.M. 1995. Rapid build-up of fish biomass in a Caribbean marine reserve. *Conservation Biology* 9:815-826.

Roberts, C.M. 1996. Settlement and beyond: population regulation and community structure. Pp. 85-112 in N.V.C. Polunin and C.M. Roberts (eds.), *Reef Fisheries*, Chapman and Hall. London.

Roberts, C.M. 1997a. Ecological advice for the global fisheries crisis. *Trends in Ecology and Evolution* 12:35-38.

Roberts, C.M. 1997b. Connectivity and management of Caribbean coral reefs. *Science* 278:1454-1457.

Roberts, C.M. 1998a. Permanent no-take zones: A minimum standard for effective marine protected areas. Pp. 96-100 in M.E. Hatziolos, A.J. Hooten, and M. Fodor (eds.), *Coral Reefs: Challenges and Opportunities for Sustainable Management.* The World Bank, Washington, D.C.

Roberts, C.M. 1998b. Sources, sinks and the design of marine reserve networks. *Fisheries* 23:16-19.

Roberts, C.M. 2000. Selecting marine reserve locations: Optimality versus opportunism. *Bulletin of Marine Science* 66(3): 581-592.

Roberts, C.M. In press. Marine protected areas and biodiversity conservation. In E. Norse and L. Crowder (eds.), *Marine Conservation Biology.* Island Press, Washington, D.C.

Roberts, C.M. In review a. How much of the sea should be included in fully protected marine reserves? *Ecological Applications.*

Roberts, C.M. In review b. Benefits of fully-protected marine reserves for migratory species. *Reviews in Fish Biology and Fisheries.*

Roberts, C.M., and J.P. Hawkins. 1997. How small can a marine reserve be and still be effective? *Coral Reefs* 16:150.

Roberts, C.M., and J.P. Hawkins. 1999. Extinction risk in the sea. *Trends in Ecology and Evolution* 14(6):241-246.

Roberts, C.M., and N.V.C. Polunin. 1991. Are marine reserves effective in management of reef fisheries? *Reviews in Fish Biology and Fisheries* 1:65-91.

Roberts, C.M., and N.V.C. Polunin. 1993a. Marine reserves: Simple solutions to managing complex fisheries? *Ambio* 22:363-368.

Roberts, C.M., and N.V.C. Polunin. 1993b. Effects of marine reserve protection on northern Red Sea fish populations. *Proceedings of the 7th International Coral Reef Symposium, Guam* 2:969-977.

Roberts, C.M., and N.V.C. Polunin. 1994. Hol Chan: Demonstrating that marine reserves can be remarkably effective. *Coral Reefs* 13:90.

Roberts, C.M., S. Andelman, G. Branch, R. Bustamente, J.C. Castilla, J. Dugan, B. Halpern, K. Lafferty, H. Leslie, J. Lubchenco, D. McArdle, H. Possingham, M. Ruckleshaus, and R. Warner. In review a. Ecological criteria for evaluating candidate sites for marine reserves. *Ecological Applications.*

Roberts, C.M., G. Branch, R. Bustamente, J.C. Castilla, J. Dugan, B. Halpern, K. Lafferty, H. Leslie, J. Lubchenco, D. McArdle, M. Ruckleshaus, and R. Warner. In review b. Application of ecological criteria in selecting marine reserves and developing reserve networks. *Ecological Applications.*

Robinson, M. 1999. *The Status of Washington's Coastal Marine Protected Areas.* Washington Department of Fish and Wildlife, Interjurisdictional Resource Management, Olympia.

Rose, K. 2000. Why are quantitative relationships between environmental quality and fish populations so elusive. *Ecological Applications* 10(2):367-385.

Rosenberg, A., T.E. Bigford, S. Leathery, R.L. Hill, and K. Bickers. 2000. Ecosystem approaches to fishery management through essential fish habitat. *Bulletin of Marine Science* 66(3):535-542.

Roughgarden, J. 1998. How to manage fisheries. *Ecological Applications* 8:S160-S164.

Rounsefell, G.A. 1975. *Ecology, Utilization, and Management of Marine Fisheries*. C.V. Mosby, Saint Louis, MO.

Rowley, R.J. 1994. Case studies and reviews: Marine reserves in fisheries management. *Aquatic Conservation: Marine and Freshwater Ecosystems* 4:233-254.

Ruckelshaus, M.H., and C.G. Hays. 1998. Conservation and management of species in the sea. Pp. 110-156 in P.L. Fiedler and P.M. Karieva (eds.), *Conservation Biology for the Coming Decade*. Chapman and Hall, London.

Russ, G. 1985. Effects of protective management on coral reef fishes in the central Philippines. *Proceedings of the 5th International Coral Reef Symposium, Tahiti*, 4:219-224.

Russ, G.R., and A.C. Alcala. 1989. Effects of intense fishing pressure on an assemblage of coral reef fishes. *Marine Ecology Progress Series* 56:13-28.

Russ, G.R., and A.C. Alcala. 1994. Sumilon Island reserve: 20 years of hopes and frustrations. *Naga, the ICLARM Quarterly* 7:8-12.

Russ, G.R., and A.C. Alcala. 1996. Marine reserves: Rates and patterns of recovery and decline of large predatory fish. *Ecological Applications* 6(3):947-961.

Sadovy, Y. 1993. The Nassau grouper, endangered or just unlucky? *Reef Encounter* 13:10-12.

Sadovy, Y. 1994. Grouper stocks of the western central Atlantic: The need for management and management needs. *Proceedings of the Gulf Caribbean Fisheries Institute* 43:43-64.

Sadovy, Y., and A.E. Eklund. 2000. Synopsis of biological data on the Nassau grouper, *Epinephelus striatus* (Bloch, 1792), and the jewfish, *E. itajara* (Lichtenstein, 1822). NOAA Technical Report NMFS 146. FAO Fisheries Synopsis 157. National Marine Fisheries Service, Seattle.

Safina, C. 1998a. Scorched-earth fishing. *Issues in Science and Technology* 14:33-36.

Safina, C. 1998b. *Song for the Blue Ocean*. Henry Holt, New York.

Sagarin, R.D., J.P. Barry, S.E. Gilman, and C.H. Baxter. 1999. Climate-related changes in an intertidal community over short and long time scales. *Ecological Monographs* 69:465-490.

Salm, R.V., and J.R. Clark. 1984. *Marine and Coastal Protected Areas: A Guide for Planners and Managers*. International Union for the Conservation of Nature and Natural Resources, Gland, Switzerland.

Salm, R.V., and A. Price. 1995. Selection of marine protected areas. Pp. 15-31 in S. Gubbay (ed.), *Marine Protected Areas: Principles and Techniques for Management*. Chapman and Hall, London.

Salm, R.V., and J.R. Clark (with E. Siirila). 2000. *Marine and Coastal Protected Areas: A Guide for Planners and Managers*, 3rd Edition. The World Conservation Union (IUCN), Gland, Switzerland.

Sammarco, P.W., and J.C. Andrews. 1988. Localized dispersal and recruitment in Great Barrier Reef corals: The Helix Experiment. *Science* 239:1422-1424.

Samuelson, P.A. 1954. The pure theory of public expenditure. *Review of Economics and Statistics* 37:350-356.

San Juan County Marine Resources Committee. Undated. *Bottomfish Recovery Program brochure*. San Juan County Marine Resources Committee, Friday Harbor, Wash.

Scanlon, K.M. 1998. *Oculina Bank: Geology of a Deep-Water Coral Reef Habitat off Florida*. U.S. Geological Survey Information Handout. U.S. Department of the Interior, Washington, D.C.

Scheltema, R.S. 1986. Long distance dispersal by planktonic larvae of shoal-water benthic invertebrates among central Pacific islands. *Bulletin of Marine Science* 39:241-256.

Schirripa, M.J., and C.M. Legault. 1999. Status of the red snapper in U.S. waters of the Gulf of Mexico: Updated through 1998. *Sustainable Fisheries Division contribution SFD-99/00-75*. National Marine Fisheries Service, Southeast Fisheries Science Center, Miami.

Schmidt, K.F. 1997. No-take zones spark fisheries debate. *Science* 277:489-491.

Seaborn, C. 1996. *Underwater Wilderness: Life in America's National Marine Sanctuaries and Reserves.* Roberts Rinehart Publishers, Boulder, Colo.

Secor, D.H., E.D. Houde, and D.M. Monteleone. 1995. A mark-release experiment on larval striped bass, *Morone saxatilis,* in a Chesapeake Bay tributary. *ICES Journal of Marine Science* 52:87-101.

Shanks, A., and B. Grantham. In review. Propagule dispersal distance and the size and spacing of marine reserves. *Ecological Applications*

Shapiro, D.Y. 1978. Reproduction in groupers. Pp. 295-327 in J.J. Polovina and S. Ralston (eds.), *Tropical Snappers and Groupers: Biology and Fisheries Management.* Westview Press, Boulder, Colo.

Shapiro, D.Y. 1979. Social behavior, group structure, and the control of sex reversal in hermaphroditic fish. *Advanced Study of Behavior* 10:43-102.

Shea, K., P. Amarasekare, P. Kareviva, M. Mangel, J. Moore, W. Murdoch, E. Noonburg, M.A. Pascual, A.M. Parma, H.P. Possingham, C. Wilcox, and D. Yu. 1998. Population management in conservation, harvesting and control. *Trends in Ecology and Evolution* 13:371-375.

Shepherd, J. G. 1988. Fish stock assessments and their data requirements. Pp. 35-62 in Gulland, J. A. (ed.). *Fish Population Dynamics,* second edition. J. Wiley & Sons, Chichester.

Shulman, M.J., and E. Bermingham. 1995. Early life histories, ocean currents, and the population genetics of Caribbean reef fishes. *Evolution* 49: 897-910.

Simberloff, D. 1988. The contribution of population and community biology to conservation science. *Annual Review of Ecological Systems* 19:473-511.

Simberloff, D. 2000. No reserve is an island: Marine reserves and nonindigenous species. *Bulletin of Marine Science* 66(3):567-580.

Sinclaire, A.D., D. Gascon, R. O'Boyle, D. Rivard, and S. Gavaris. 1991. Consistency of some southwest Atlantic groundfish stock assessments. *Northwest Atlantic Fisheries Organization Scientific Council Studies* 16:59-77.

Sladek Nowlis, J.S. 2000. Short- and long-term effects of three fishery-management tools on depleted fisheries. *Bulletin of Marine Science 66(3): 651-662.*

Sladek Nowlis, J.S., and C.M. Roberts. 1997. You can have your fish and eat it too: Theoretical approaches to marine reserve design. *Proceedings of the 8th International Coral Reef Symposium* 2:1907-1910.

Sladek Nowlis, J.S., and C.M. Roberts. 1999. Fisheries benefits and optimal design of marine reserves. *Fisheries Bulletins U.S.* 97.

Sladek Nowlis, J.S. and M.M. Yoklavich. 1998. Design criteria for rockfish harvest refugia from models of fish transport. Pp. 32-40 in M.M. Yoklavich (ed.), *Marine Harvest Refugia for West Coast Rockfish: A Workshop.* NOAA Technical Memorandum. NOAA-TM-NMFS-SWFSC-225. U.S. Department of Commerce, Pacific Grove, CA.

Sluka, R., M. Chiappone, K.M. Sullivan, and R. Wright. 1997. The benefits of a marine fishery reserve for Nassau grouper *Epinephelus striatus* in the central Bahamas. *Proceedings of the 8th International Coral Reef Symposium, Panama* 2:1961-1964.

Smith, A.H., and F. Berkes. 1991. Solutions to the "tragedy of the commons": Sea urchin management in St. Lucia, West Indies. *Environmental Conservation* 18(2):131-136.

Soh, S.K., D.R. Gunderson, and D.H. Ito. 1998. Closed areas to manage rockfishes in the Gulf of Alaska. Pp. 118-124 in M.M. Yoklavich (ed.), *Marine Harvest Refugia for West Coast Rockfish: A Workshop.* NOAA Technical Memorandum NOAA-TM-NMFS-SWFSC-255. U.S. Department of Commerce, Pacific Grove, CA.

Starrett, D. 1988. *Foundations of Public Economics.* Cambridge University Press, Cambridge, U.K.

Steele, J.H. 1985. A comparison of terrestrial and marine ecological systems. *Nature* 313:355-358.

Steele, J.H. 1991. Marine ecosystem dynamics: Comparison of scales. *Ecological Research* 6:175-183.

Steele, J.H. 1996. Regime shifts in fisheries management. *Fisheries Research* 25:19-23.

Stoffle, R., F. Jensen, and D. Rasch. 1987. Cultural basis of sport anglers' response to reduced lake trout catch limits. *Transactions of the American Fisheries Society* 116:503-509.

Stoffle, R., D. Halmo, T. Wagner, and J. Luczkovich. 1994a. Reefs from space: Satellite imagery, marine ecology, and ethnography in the Dominican Republic. *Human Ecology* 22(3):355-378.

Stoffle, B., R. Stoffle, D. Halmo, and C. Burpee. 1994b. Folk management and conservation ethics among small-scale fishermen of Buren Hombre, Dominican Republic. Pp. 115-138 in C. Dyer and J. McGoodwin (eds.), *Folk Management in the World's Fisheries*. University of Colorado Press, Boulder.

Stoffle, R., D. Austin, and D. Halmo. 1997. Cultural landscapes and traditional cultural properties: A Southern Paiute view of the Grand Canyon and Colorado River. *American Indian Quarterly* 21(2):229-249.

Stokes, T.K. 1997. Review of red snapper (*Lutjanus campechanus*) research and management in the Gulf of Mexico. Science and Management Review Panel, submitted to NOAA/NMFS September 1997. National Oceanic and Atmospheric Administration, Silver Spring, MD.

Stoner, D.S. 1992. Vertical distribution of a colonial ascidian on a coral reef: the roles of larval dispersal and life-history variation. *American Naturalist* 139(4):802.

Stoner, A.W., and M. Ray. 1996. Queen conch, *Strombus gigas*, in fished and unfished locations of the Bahamas: Effects of a marine fishery reserve on adults, juveniles, and larval production. *Fishery Bulletin* 94:551-565.

Sumaila, U.R. 1998. Protected marine reserves as fisheries management tools: A bioeconomic analysis. *Fantoftvegen* 38:N-5036.

Suman, D. 1997. The Florida Keys National Marine Sanctuary: A case study of an innovative federal-state partnership in marine resource management. *Coastal Management* 25(3):293-324.

Suman, D. 1998. Stakeholder group perceptions of marine reserves in the Florida Keys National Marine Sanctuary. Pp. 100-112 in M.M. Yoklavich (ed.), *Marine Harvest Refugia for West Coast Rockfish: A Workshop*. NOAA Technical Memorandum. NOAA-TM-NMFS-SWFSC-225. U.S. Department of Commerce, Pacific Grove, CA.

Suman, D., M. Shivlani, and J.W. Milon. 1999. Perceptions and attitudes regarding marine reserves: A comparison of stakeholder groups in the Florida Keys National Marine Sanctuary. *Ocean and Coastal Management* 42:1019-1040.

Swearer, S.E., J. Caselle, D. Lea, and R.R. Warner. 1999. Larval retention and recruitment in an island population of a coral-reef fish. *Nature* 402:799-802.

Tegner, M. J. 1992. Brood-stock transplants as a approach to abalone stock enhancement. Pp. 608 in S.A. Shepherd, M.J. Tegner, and S.A. Guzman del Proo (eds.), *Abalone of the World: Their Biology, Fisheries and Culture*. Blackwell Scientific Publications, Oxford, U.K.

Tegner, M.J. 1993. Southern California abalones: Can stocks be rebuilt using marine harvest refugia? *Canadian Journal of Fisheries and Aquatic Science* 50:2010-2018.

Tegner, M.J., L.V. Basch, and P.K. Dayton. 1996. Near extinction of an exploited marine invertebrate. *Trends in Ecology and Evolution*: 11: 278-280.

Thorne-Miller, B., and J. Catena. 1991. *The Living Ocean: Understanding and Protecting Marine Biodiversity*. Island Press, Washington, D.C.

Thorpe, J.E., R.I.G. Morgan, C. Talbot, and M.S. Miles. 1983. The inheritance of developmental rates in Atlantic salmon, Salmo salar L. *Aquaculture* 33: 119-128.

Thrush, S.F., J.E. Hewitt, V.J. Cummings, P.K. Dayton, M. Cryer, S.J. Turner, G.A. Funnel, R.G. Budd, C.J. Milburn, and M.R. Wilkinson. 1998. Disturbance of the marine benthic habitat by commercial fishing: impacts at the scale of the fishery. *Ecological Applications* 8: 866-879.

Ticco, Paul. 1996. The use of marine protected areas to preserve and enhance marine biological diversity: A case study approach. *Coastal Management* 24(3):309-314.

Trexler, J., and J. Travis. 2000. Can marine protected areas restore and conserve stock attributes of reef fishes? *Bulletin of Marine Science* 66(3):853-873.

Tuan, Yi-Fu. 1996. *Place and Space: The Perspective of Experience.* University of Minnesota Press, Minneapolis.

Turpie, J.K., L.E. Beckley, and S.M. Katua. 2000. Biogeography and the selection of priority areas for conservation of South African coastal fishes. *Biological Conservation* 92:59-72.

United Nations (UN). 1983. *Law of the Sea.* St. Martin's Press, Inc., New York (also available on-line at http://fletcher.tufts.edu/multi/texts/BH825.txt).

United Nations Educational, Scientific and Cultural Organization (UNESCO). 1984. *Action Plan for Biosphere Reserves.* UNESCO, Paris, France.

United Nations Environment Programme (UNEP). 1988. *Environmental Perspective to the Year 2000 and Beyond.* United Nations Environment Programme, Nairobi, Kenya.

United Nations Environment Programme (UNEP). 1992. *Convention on Biological Diversity.* United Nations Environment Programme, Nairobi, Kenya (also available on-line at http://biodiv.org/chm/conv/default.htm).

U.S. Coral Reef Task Force (USCRTF). 2000. *National Action Plan for Coral Reef Conservation.* U.S. Coral Reef Task Force, Washington, D.C. (also available on-line at http://coralreef.gov/WG-reports.html).

U.S. Department of Agriculture, Forest Service. 1993. *The Principal Laws Relating to Forest Service Activities.* U.S. Government Printing Office, Washington, D.C.

U.S. Department of Commerce (DOC). 1994. *Florida Keys National Marine Sanctuary Draft Environmental Impact Statement/Management Plan. Volume 1. Development of the Management Plan: Environmental Impact Statement.* Sanctuaries and Reserves Division, National Oceanic and Atmospheric Administration, Silver Spring, MD.

Wallace, S.S. 1999. Evaluating the effects of three forms of marine reserve on northern abalone populations in British Columbia, Canada. *Conservation Biology* 13:882-887.

Walls, K. 1998. Leigh Marine Reserve, New Zealand. *Parks* 8:5-10.

Walters, C.J. 1998. Designing fisheries management systems that do not depend on accurate stock assessment. Pp. 279-288 in T.J. Pitcher, P.J.B. Hart, and D. Pauly (eds.), *Reinventing Fisheries Management.* Chapman and Hall, London.

Walters, C.J. 2000. Impacts of dispersal, ecological interactions, and fishing effort dynamics on efficacy of marine protected areas: How large should protected areas be? *Bulletin of Marine Science* 66(3):745-757.

Walters, C.J., V. Christensen, and D. Pauly. 1997. Structuring dynamic models of exploited ecosystems from trophic mass-balance assessments. *Review of Fish Biology and Fisheries.* 7:139-172.

Walters, C., and J-J. Maguire. 1996. Lessons for stock assessment from the northern cod collapse. *Review of Fish Biology and Fisheries* 6:145-159.

Walters, C., and A.M. Parma. 1996. Fixed exploitation rate strategies for coping with effects of climate change. *Canadian Journal of Fisheries and Aquatic Sciences* 53:148-158.

Walters, C., and P.H. Pearse. 1996. Stock information requirements for quota management systems in commercial fisheries. *Review of Fish Biology and Fisheries* 6:21-42.

Wantiez, L., P. Thollot, and M. Kulbicki. 1997. Effects of marine reserves on coral reef fish communities from five islands in New Caledonia. *Coral Reefs* 16:215-224.

Ward, T.J., M.A. Vanderklift, A.O. Nicholls, and R.A. Kenchington. 1999. Selecting marine reserves using habitats and species assemblages as surrogates for biological diversity. *Ecological Applications* 9:691-698.

Warner, R.R., S. Swearer, and J.E. Caselle. 2000. Larval accumulation and retention: Implications for the design of marine reserves and essential fish habitat. *Bulletin of Marine Science* 66(3): 821-830

Washington Department of Natural Resources. 1996. *Cypress Island Natural Resources Conservation Area: Management Plan.* Washington Department of Natural Resources, Olympia.

Watling, L., and E.A. Norse. 1998. Disturbance of the seabed by mobile fishing gear: A comparison to forest clearcutting. *Conservation Biology* 12:1180-1197.

Watson, M., and J.L. Munro. In press. Maximizing settlement success in depleted marine reserves. *Bulletin of Marine Science.*

Watson, M., and R.F.G. Ormond. 1994. Effect of an artisanal fishery on the fish and urchin populations of a Kenyan coral reef. *Marine Ecology Progress Series* 109(2-3):115-129.

Watson M., D. Righton, T. Austin, and R. Ormond. 1996. The effects of fishing on coral reef fish abundance and diversity. *Journal of the Marine Biological Association of the United Kingdom* 76(1):229-233.

Watson, M., R.F.G. Ormond, and L. Holliday. 1997. The role of Kenya's marine protected areas in artisanal fisheries management. *Proceedings of the 8th International Coral Reef Symposium, Panama* 2:1955-1960.

Wells, S. 1997. Giant clams: Status, trade and mariculture, and the role of CITES in management. Species Survival Commission, International Union for the Conservation of Nature and Natural Resources, Gland, Switzerland.

Wenner, E.L., and M. Geist. 2001. The National Estuarine Research Reserve's program to monitor and preserve estuarine waters. *Coastal Management* 29:1-18.

Wilcove, D.S., and E.O. Wilson. 2000. *The Condor's Shadow: The Loss and Recovery of Wildlife in America.* Bantam-Doubleday-Dell, New York.

Wilcove, D.S., D. Rothstein, J. Dubow, A. Phillips, and E. Losos. 1998. Quantifying threats to imperiled species in the United States. *BioScience* 48: 607-615.

Wilkinson, C.F., and H.M. Anderson. 1987. *Land and Resource Planning in the National Forests.* Island Press, Washington, D.C.

Williams, W. 1980. *The Implementation Perspective.* University of California Press, Berkeley.

Willig, R. 1976. Consumers' surplus without apology. *American Economic Review* 66:589-597.

Willison, M., D.P. Jones, and S. Atwood. In press. Deep sea corals and marine protected areas in Nova Scotia. *Proceedings of the 4th International Conference on Science and the Management of Protected Areas* (SAMPAA), Wolfville, Canada.

Wilson, C. A., J. H. Render, D. L. Neiland. 1994. Life history gaps in red snapper (*Lutjanus campechanus*), swordfish (*Xiphias gladius*), and red drum (*Scianeops ocellatus*) in the northern Gulf of Mexico: Age distribution, growth, and some reproductive biology. *Final Report Cooperative Agreement NA17FF0383-02.* U.S. Department of Commerce, National Marine Fisheries Service, Marine Fisheries Institute.

Witherell, D., C. Pautzke, and D. Fluharty. 2000. Integrating ecosystem considerations into groundfish fisheries management off Alaska, USA. *International Council for the Exploration of the Sea.*

Wolverton, R., and W. Wolverton. 1994. *The National Seashores: The Complete Guide to America's Scenic Coastal Parks.* Roberts-Rinehart Publishers, Niwot, Colo.

World Commission on Environment and Development (WCED). 1987. *Our Common Future.* Oxford University Press, Oxford, U.K.

Zacharias, M., D. Howes, J. Harper, and P. Wainwright. 1998. The British Columbia marine ecosystem classification: Rationale, development, and verification. *Coastal Management* 26:105-124.

Zedeno, N., D. Austin, R. Stoffle. 1997. Landmark and landscape: A contextual approach to the management of American Indian resources. *Culture and Agriculture* 19:123-129.

Zeller, D.C., and G.R. Russ. 1998. Marine reserves: Patterns of adult movement of the coral trout (*Plectropomus leopardus* (Serranidae)). *Canadian Journal of Fisheries and Aquatic Science* 55:917-924.

APPENDIXES

A

Acronyms

AFS	American Fisheries Society
CBD	Convention on Biological Diversity
CBRA	Coastal Barrier Resources Act
CERCLA	Comprehensive Environmental Response, Compensation and Liability Act
CMC	Center for Marine Conservation
CNPPA	Commission on National Parks and Protected Areas (IUCN)
COMPARE	Criteria and Objectives for Marine Protected Area Evaluation
CPUE	Catch per unit effort
CVM	Contingent value method
CWA	Clean Water Act
CZMA	Coastal Zone Management Act
DOC	U.S. Department of Commerce
EEZ	Exclusive economic zone
EFH	Essential fish habitat
ESA	Endangered Species Act
FAO	Food and Agriculture Organization
FKNMS	Florida Keys National Marine Sanctuary
FWCA	Fish and Wildlife Coordination Act

FWS   Fish and Wildlife Service

GESAMP   Joint Group of Experts on the Scientific Aspects of Marine
                 Pollution
GIS   Geographical information system
GPS   Global positioning system

HA   Hedonic approach
HAPC   Habitat area of particular concern

IFQ   Individual Fishing Quota
IMO   International Maritime Organization
ITC   Individual transferable quota
IUCN   International Union for the Conservation of Nature and Natural
                 Resources

MARPOL   International Convention for the Prevention of Pollution from
                 Ships
MMA   Marine managed area
MMPA   Marine Mammals Protection Act
MPA   Marine protected area
MPPRC   Marine Plastics Pollution Research and Control Act
MSFCMA   Magnuson-Stevens Fishery Conservation and Management Act
MSY   Maximum sustainable yield

NAPA   National Academy of Public Administration
NEPA   National Environmental Policy Act
NERR   National Estuarine Research Reserve
NERRS   National Estuarine Research Reserve System
NMFS   National Marine Fisheries Service
NMS   National Marine Sanctuary
NMSP   National Marine Sanctuary Program
NMSA   National Marine Sanctuary Act
NOAA   National Oceanic and Atmospheric Administration
NOS   National Ocean Service
NPS   National Park Service
NRC   National Research Council
NWR   National Wildlife Refuge
NPFMC   North Pacific Fisheries Management Council

OCS   Outer Continental Shelf
OCSLA   Outer Continental Shelf Lands Act
OPA   Oil Pollution Act

RFPDT   Reef Fishery Plan Development Team

SPAW   Protocol on Specially Protected Areas and Wildlife of the
                 Wider Caribbean Region
SWNP   System-Wide Monitoring Program

TAC   Total allowable catch
TCM   Travel-cost model
TNC   The Nature Conservancy
TPA   Terrestrial protected area
TURF   Territorial use right fishery

UN   United Nations
UNCLOS   United Nations Convention on the Law of the Sea
UNEP   United Nations Environment Programme
UNESCO   United Nations Educational, Scientific, and Cultural Organiza-
                 tion
USCRTF   U.S. Coral Reef Task Force
USFS   U.S. Forest Service

VMS   Vessel monitoring system

WCED   World Commission on Environment and Development
WCPA   World Commission on Protected Areas (IUCN)
WWF   World Wildlife Fund

B

Glossary

Allee effect: A reduction in fitness at low population densities, often measured as the numbers of offspring that are produced or survive. For example, many marine species reproduce by releasing eggs and sperm into the water where they are fertilized externally. The rate of fertilization is greatly reduced as the distance between reproductive partners increases. For animals that have low mobility, such as clams that are attached to the seabed, reductions in population density can prevent effective reproduction long before all the individuals have been removed. Strong Allee effects render populations vulnerable to extinction when their densities have been reduced to low levels, for example, by fishing. They also hinder the recovery of populations from low densities.

biodiversity: The variation in living systems at all organizational levels, from the large-scale diversity of ecosystems to the minutiae of genetic diversity within a particular population. It is often evaluated through measurement of species diversity in a given area or over a specified period of time.

biota: The plant and animal life characteristic of a specific region, or biosphere, or given time period.

buffer zone: The area that separates the core from areas in which human activities that threaten it occur.

co-management: Management carried out by government and local communities in partnership.

connectivity: The movement of organisms from place to place (e.g., among reserves) through dispersal or migration.

core area: The central, most highly protected part of a protected area.

critical areas: Areas within an MPA that are crucial to achieving the objectives of the MPA; for example, spawning areas in an MPA established for fisheries purposes.

cultural landscape: A cluster of beliefs, values, and norms about how are places and things on the earth are related to human behavior.

culturally affiliated: To be connected to a place, region, or resource because it has significant meaning in the culture of the individual and his or her group. In most cases, cultural affiliation requires more than one generation to establish, and for some groups the connections have been developed over centuries. The federal government uses the term in various environmental laws and regulations.

ecological reserve: Zoning that protects all living marine resources through prohibitions on fishing and on the removal or disturbance of any living or nonliving marine resource. Access and recreational activities may be restricted to prevent damage to the resources. These reserves may also be referred to as fully protected areas.

ecosystem: An integrated system of living species, their habitat, and the processes that affect them.

ecosystem approach: Fishery management actions aimed at conserving the structure and function of marine ecosystems, in addition to conserving the fishery resource.

endemism: Of or relating to a native species or population occurring under highly restricted conditions due to the presence of a unique environmental factor that limits its distribution.

environmental ethics: A cluster of beliefs, values, and norms regarding how humans should interact with the environment.

exclusive economic zone (EEZ): All waters from the seaward boundary of coastal nations to 200 nautical miles.

existence value: see heritage value.

fishery reserve: Zoning that precludes fishing activity on some or all species to protect critical habitat, rebuild stocks (long term, but not necessarily permanent closure), provide insurance against overfishing, or enhance fishery yield.

growth overfishing: Fishing mortality at which the losses in weight from total mortality exceed the gain in weight due to growth. Growth overfishing results from catching too many small fish before they have reached an optimum marketable size.

heritage (or existence) value: Site possessing historical, archaeological, architectural, technological, aesthetic, scientific, spiritual, social, traditional, or other special cultural significance associated with human activity.

individual fishing quota (IFQ): Fishery management tool used in the Alaska halibut and sablefish, wreckfish, and surf clam and ocean quahog fisheries in the United States, and other fisheries throughout the world, that allocates a certain portion of the total allowable catch to individual vessels, fishermen, or other eligible recipients based on initial qualifying criteria.

individual transferable quota (ITQ): Individual fishing quota that can be transferred. ITQs typically entail allocations of a certain amount of an established annual catch to individual fishermen or vessel owners. Once distributed, fishermen can buy or sell their share, or individual quota, to other fishermen or vessel owners.

integrated management: An approach by which the many competing environmental and socioeconomic issues are considered together, with the aim of achieving the optimal solution from the viewpoint of the whole community and the whole ecosystem.

marine protected area (MPA): Geographic area with discrete boundaries that has been designated to enhance the conservation of marine resources. This includes MPA-wide restrictions on some activities such as oil and gas mining and the use of zones such as fishery and ecological reserves to provide higher levels of protection.

marine reserve: A zone in which some or all of the biological resources are protected from removal or disturbance; encompasses both fishery and ecological reserves.

maximum sustainable yield (MSY): Largest average catch that can be harvested on a sustainable basis from a stock under existing environmental conditions. MSY is a deterministic single-species construct that may have difficulty reflecting the stochastic nature of stock dynamics.

metapopulation: A population that consists of a series of physically separate subpopulations linked by dispersal. Metapopulations persist as a result of a balance between extinctions of subpopulations and recolonization of habitat patches (and hence reestablishment of subpopulations).

monitoring system: A system designed to reveal the extent to which the ecological and socioeconomic objectives of an MPA are being met, as a basis for management actions.

multiple-use MPA: An approach, often employed over much larger areas, that allows for integrated management of complete marine ecosystems, usually through a zoning process.

National Marine Fisheries Service (NMFS): Federal agency within NOAA responsible for overseeing fisheries science and regulation.

network: A group of reserves designed to meet objectives that single reserves cannot achieve on their own. Networks of reserves are linked by dispersal of marine organisms and by ocean currents.

open access: A type of fishery in which anyone wanting to fish who has the appropriate gear can do so. A fishery is considered open access even when licenses are required, if the number of licenses is not limited and the holder does not have to abide by individual quotas or other restrictions to access.

precautionary approach: a management philosophy that favors constraining an activity when there is high scientific uncertainty regarding its effects on the natural environment, as opposed to allowing an activity to continue until proof, of either no effect or a negative impact, is obtained.

protected area: An area of land and/or sea especially dedicated to the protection and maintenance of biological diversity and of natural and associated cultural resources, and managed through legal or other effective means.

recruitment: A measure of the number of fish that enter a class during some time period, such as the spawning class or fishing-size class.

recruitment overfishing: This condition results from fishing at a high enough level to reduce the biomass of reproductively mature fish (spawning biomass) to a level at which future recruitment is reduced. Recruitment overfishing is characterized by a decreasing proportion of older fish in the fishery and consistently low average recruitment over time.

regional fishery management councils: Eight regional councils mandated in the Magnuson-Stevens Fishery Conservation and Management Act to develop management plans for fisheries in federal waters.

sectoral management: A management approach in which specific agencies are given responsibility for managing particular sectors. Examples are fishery management agencies and tourism management agencies. The result of *sectoral* management of an area in which different sectors compete for resources is often conflict between users, and between different sector management agencies with responsibilities over a common area, even under the same government. There is an inherent incentive for each sector to maximize its profits and benefits at the expense of other sectors, the general public or the natural environment.

sink: Habitats in which birth rates are lower than death rates and emigration is lower than immigration, as applied to equilibrium populations. A more general definition is that a sink is a compartment that is a net importer of individuals.

source: Patches in which birth rates are higher than death rates and emigration rates are higher than immigration rates, as applied to equilibrium populations. A more general defintion is a compartment that, over a large period of time (e.g., several generations), shows no net change in population size but nonetheless is a net exporter of individuals.

spawning stock biomass: The total weight of mature fish in a stock, often expressed in relative terms, i.e., as a percentage of the mature biomass when fishing mortality is zero.

stakeholders: Refers to anyone who has an interest in or who is affected by the establishment of a protected area.

sustainability: The use of ecosystems and their resources in a manner that satisfies current needs without compromising the needs or options of future generations.

total allowable catch (TAC): The annual recommended catch for a species or species group. The regional council sets the TAC from the range of allowable biological catch.

zoning: A process in which a protected area is divided into discrete zones and particular human uses of each zone are permitted, often with conditions such as gear limitations in fishing and waste discharge prohibitions in tourism.

C

Committee and Staff Biographies

COMMITTEE MEMBERS

Dr. Edward Houde, chair of this committee, earned his Ph.D. in fishery science from Cornell University in 1968. Dr. Houde is currently a professor in the University of Maryland's Center for Environmental Science. His research interests include fisheries science and management, larval fish ecology, fisheries oceanography, and aquatic resources management. Dr. Houde has served previously as Director of NSF's Biological Oceanography Program. He is the recipient of the Beverton (Fisheries Society of the British Isles) and Sette (American Fisheries Society) Awards for career achievement. Dr. Houde is a member of the Ocean Studies Board and has served on numerous advisory committees, including the International Council for the Exploration of the Sea, the NRC Committee on Sustaining Marine Fisheries, and the NMFS Ecosystem Principles Advisory Panel.

Dr. Felicia C. Coleman earned her Ph.D. in biological science from Florida State University in 1991. Dr. Coleman is currently an associate in research at Florida State University. Her research interests focus on reef fish, particularly their population ecology and the effects of fishing on reproduction. She organized the Mote International Symposium on Essential Fish Habitat and Marine Reserves (1998) and was guest editor for publication of the proceedings in the *Bulletin of Marine Science* (May 2000). She recently was named an Aldo Leopold Conservation Leadership Fellow by the Ecological Society of America.

Dr. Paul Dayton earned his Ph.D. in zoology from the University of Washington in 1970. Dr. Dayton is currently a professor at the Scripps Institution of Oceanography. His primary research interests include coastal ecology with a recent interest in the impacts of fishing on coastal ecosystems. Dr. Dayton received the George Mercer Award from the Ecological Society of America in 1974 and is a fellow of the American Association for the Advancement of Science. Dr. Dayton received a Pew Fellowship for Marine Conservation Research in 1994.

Dr. David Fluharty earned his Ph.D. in natural resources conservation and planning from the University of Washington in 1976. Dr. Fluharty is currently a professor at the University of Washington in Seattle, where he teaches a course on management of marine protected areas. His research interests include natural resources policy at national and international levels, and management of marine resources, particularly fisheries. He currently is a voting member of the North Pacific Fishery Management Council. Recently, he served as chair of the NMFS Ecosystem Principles Advisory Panel.

Mr. Graeme Kelleher earned a B.E. in civil engineering from the University of Sydney in 1955. Mr. Kelleher is currently a Consultant for the Great Barrier Reef Marine Park Authority of which he was the chair and chief executive from 1979 to 1994. His research interests include establishment and management of marine protected areas and application of the concept of ecologically sustainable development to the management of large marine ecosystems. He coauthored "Guidelines for Establishing Marine Protected Areas" (IUCN, 1992) and received the Fred M. Packard International Parks Merit Award from the IUCN in 1998.

Dr. Stephen Palumbi earned his Ph.D. in zoology from the University of Washington in 1984. Dr. Palumbi is currently a professor at Harvard University. His research interests include speciation mechanisms in marine systems, population structure of species with high dispersal potential, and population genetics of source and sink populations of marine invertebrates and mammals. He is a co-director of the National Center for Ecological Analysis and Synthesis Program in Developing the Theory of Marine Reserves. In 1996, Dr. Palumbi received a Pew Fellowship for Marine Conservation Research.

Dr. Ana Maria Parma earned her Ph.D. in fisheries science from the University of Washington in 1988. Dr. Parma is currently a population dynamicist at the Centro Nacional Patagonico in Argentina. Her research interests include fish stock assessment, population dynamics, analysis of stochastic models, and adaptive management of fisheries resources. Dr. Parma was awarded the P.E.O. International Peace Scholarship in 1985.

Dr. Stuart Pimm earned his Ph.D. in biology from New Mexico State University in 1974. Dr. Pimm is currently a professor at the Center for Environmental Research and Conservation at Columbia University. His research interests include determination of the extinction rates of birds and other animals, conservation biology, ecology, and evolutionary biology. Dr. Pimm also is working on improving the management of the Everglades National Park to preserve threatened species and the ecosystems on which they depend. In 1993, he was awarded a Pew Scholarship in Conservation and the Environment.

Dr. Callum Roberts earned his Ph.D. in biology from the University of York, United Kingdom, in 1986, where he is currently a professor. His research interests include marine conservation biology, behavior and ecology of fish on Red Sea coral reefs, management and conservation of coral reefs, origin and maintenance of biodiversity in reefs, and effects of fishing and recreational tourism on ecosystems. Currently he is working to develop design principles for effective international networks of marine protected areas. He is a member of the U.K. Steering Group on Marine Reserves.

Dr. Sharon Smith earned her Ph.D. in zoology from Duke University in 1975. Dr. Smith is currently a professor of marine biology and fisheries at the University of Miami, Rosenstiel School of Marine and Atmospheric Science. Her research interests include ecology of zooplankton, nutrient cycling, upwelling ecosystems, high latitude ecosystems, and population dynamics. She is a member of numerous advisory and steering committees, including service as the chair of the Advisory Subcommittee of the Ocean Sciences Division and as a member of the Committee on Global Ecosystem Dynamics, National Science Foundation (NSF). In addition, she is a former member of the Ocean Studies Board.

Dr. George Somero earned his Ph.D. in biological sciences from Stanford University in 1967. Dr. Somero is currently the David and Lucile Packard Professor of Marine Science at Stanford University. He studies the adaptations of organisms to marine environments, including biochemical and physiological adaptation, within the context of the biogeography and evolution of marine species. Dr. Somero is a member of the National Academy of Sciences and a fellow of the American Association for the Advancement of Science.

Dr. Richard Stoffle earned his Ph.D. in anthropology from the University of Kentucky in 1972. He is currently a senior research anthropologist at the University of Arizona. His research interests include cultural anthropology, social impact assessment, developmental anthropology, Native Americans, and the ethnography of fisheries. He has conducted a variety of studies in the Caribbean islands of Antigua and the Dominican Republic on the environmental effects of fishing behavior and conservation values of small-scale coastal fishermen. He was appointed to the Board of Technical Experts of the Great Lakes Fishery Commission in 1986 where he served for three years. He is currently conducting a study of Ojibway natural and cultural resource use in the western Great Lakes and is preparing an oral history of commercial fishermen for Isle Royale National Park.

Dr. James Wilen earned his Ph.D. in natural resource economics from the University of California, Riverside in 1973. Dr. Wilen is a professor of agriculture and resource economics at the University of California, Davis. His research is primarily in fisheries economics, with a particular focus on the analysis of alternative management strategies and institutions. He is currently involved in studies related to marine reserves, including the use of reserves as a management tool and a bioeconomic modeling project on using spatial management in the California sea urchin fishery. He has worked as a fisheries policy analyst examining a wide range of fisheries in North America and around the world.

NATIONAL RESEARCH COUNCIL STAFF

Susan Roberts (study director) received her Ph.D. in marine biology from the Scripps Institution of Oceanography. Dr. Roberts is a program officer for the National Research Council's Ocean Studies Board. Dr. Roberts staffs studies on living marine resources, marine biotechnology, and health implications of climate change. Her research interests include marine microbiology, fish physiology and development, and biomedicine.

Ann Carlisle (senior project assistant) received her B.A. in sociology from George Mason University in 1997. During her tenure with the Ocean Studies Board, she has worked on studies of the history of ocean sciences and marine living resources, and has staffed several studies on various aspects of marine fishery management.

D

Meeting Agendas

OCEAN STUDIES BOARD

Meeting of the Committee on the Evaluation, Design, and Monitoring of Marine Reserves and Protected Areas in the United States

**National Research Council, Green Building, Room 126
Washington, DC 20007**

December 14-15, 1998

Monday, Dec 14

9:30 a.m.   **Break**

OPEN SESSION

10:00 a.m.   Welcoming remarks and introduction of meeting participants
               Ed Houde, Chairman of the Committee

10:15 a.m.   Presentation of study scope and the committee's statement of task

10:30 a.m.   Introduction to marine protected areas: **Tundi Agardy**, Sr. Director Coastal Marine Conservation, Conservation International

11:15 a.m.   Jurisdictional issues in the establishment and management of marine protected areas: **Amy Browning**, University of Alaska, Fairbanks

11:30 a.m.   Committee Questions and Discussion Period

12:00 noon   **Lunch**

1:00 a.m.   Building a National System of Marine Sanctuaries: **DeWitt John**, National Academy of Public Administration

Presentations on the Use of Marine Protected Areas in New England

1:30 p.m.   Large-scale closed areas as a fishery management tool in temperate marine systems: the Georges Bank experience : **Steve Murawski**, NMFS

2:00 p.m.   Fishermen's perspective on the New England fisheries - state of the fishery, management strategies, and the use of closed areas to rejuvenate fish stocks: **Vito Calomo**, Executive Director, Gloucester Commission, et al.

2:30 p.m.   Committee Questions and Discussion Period

3:00 p.m.   **Break**

3:30 p.m.   Stellwagen Bank National Marine Sanctuary: **Bradley Barr**, Sanctuary Manager

4:00 p.m.   Ecosystem effects of fishing: **Peter Auster**

4:30 p.m.   Committee Questions and Discussion Period

5:30 p.m.   Open Meeting Adjourns for the Day

6:15 p.m.   Committee Working Dinner

Tuesday, Dec 15

OPEN SESSION

7:30 a.m.   **Breakfast**

8:30 a.m.   National Marine Fisheries Service Perspective: **William W. Fox**, Director, Office of Science and Technology

9:00 a.m.   National Ocean Service Perspective: **Stephanie Thornton**, Chief, National Marine Sanctuaries Program and National Estuarine Research Reserve System

9:30 am   Department of the Interior Perspective: **William Brown**, Science Advisor to the Secretary of the Interior

10:00 am   Review of statement of task and discussion of project goals:

                   Ed Houde, Chairman

11:00 a.m.	Open Meeting adjourns
12:00 noon	**Lunch**
1:00 p.m.	Study approach and timeline discussion continues
2:00 p.m.	Plans for next meeting and interim assignments summarized
2:30 p.m.	Meeting Adjourns

OCEAN STUDIES BOARD

Meeting of the Committee on the Evaluation, Design, and Monitoring of Marine Reserves and Protected Areas in the United States

**Cheeca Lodge
Mile Marker 82, US Highway 1
Islamorada, FL**

April 15-17, 1999

AGENDA

Thursday, April 15

8:00 a.m.	**Breakfast**
8:30 a.m.	Opening remarks and discussion of the goals of the second meeting
8:45 a.m.	Welcoming remarks and introduction of meeting participants
	Ed Houde, Chairman of the Committee Review of statement of task and progress report
9:00 a.m.	**Billy Causey** and **Ben Haskell**, Florida Keys National Marine Sanctuary
9:30 a.m.	Committee Questions and Discussion Period
9:45 a.m.	**Break**
10:00 a.m.	Roundtable with presentations from conservation groups including EDF and CMC
11:00 a.m.	Committee Questions and Discussion Period
11:15 a.m.	**Don DeMaria**, Commercial Fisherman
11:45 a.m.	Committee Questions and Discussion Period

12:00 p.m.	**Lunch**
1:30 p.m.	**Rick Ruoff**, Recreational Fishing Guide
2:00 p.m.	Committee Questions and Discussion Period
2:15 p.m.	**Daniel Suman**, University of Miami-RSMAS – Use of Marine Reserves in Coastal Management
2:45 p.m.	Committee Questions and Discussion Period
3:00 p.m.	**Russ Nelson**, Florida Marine Fisheries Commission – The State of Florida's Involvement in Marine Protected Areas
4:00 p.m.	Committee Questions and Discussion Period
4:15 p.m.	**Break**
4:30 p.m.	Open session for interested community speakers (5 min./person)
6:00 p.m.	Open Meeting Adjourns for the Day
6:30 p.m.	Committee Dinner on the beach at Cheeca Lodge

Friday, April 16

OPEN SESSION

7:30 a.m.	**Breakfast**
8:30 a.m.	**Jerry Ault**, University of Miami-RSMAS – Quantitative Assessment of Marine Reserves
9:00 a.m.	**Jim Bohnsack**, National Marine Fisheries Service – No-Take Marine Reserves: An Essential Tool for Fisheries Management
9:45 a.m.	Committee Questions and Discussion Period
10:00 a.m.	**Break**
10:15 am	**Callum Roberts** – Sources and Sinks
10:45 a.m.	Committee Questions and Discussion Period
12:00 noon	**Lunch**
1:00 p.m.	**Felicia Coleman** and **Dave Fluharty** – The Fishery Management Councils
1:30 p.m.	Question Period

1:45 p.m.	**Graeme Kelleher** – MPA nomenclature
2:15 p.m.	Committee Questions and Discussion Period
2:30 p.m.	Open Meeting Adjourns
6:00 p.m.	Meeting Adjourns for the Day

Saturday, April 17

| 7:30 a.m. | **Breakfast** |

OPEN SESSION

8:00 a.m.	**Rich Aronson**, Dauphin Island Sea Laboratory – Monitoring Closed Areas in the Florida Keys National Marine Sanctuary
8:30 a.m.	Committee question and discussion period
12:00 noon	Meeting adjourns

OCEAN STUDIES BOARD

Meeting of the Committee on the Evaluation, Design, and Monitoring of Marine Reserves and Protected Areas in the United States

**The Hilton Monterey
1000 Aguajito Road
Monterey, California**

May 27-29, 1999

AGENDA

Thursday, May 27

OPEN SESSION

11:15 a.m.	**Marc Mangel**, University of California, Santa Cruz
11:45 a.m.	Committee Questions and Discussion Period
12:00 p.m.	**Lunch**
1:00 p.m.	**Steve Webster**, Chair of the MBNMS Sanctuary Advisory Council and Senior Marine Biologist, Monterey Bay Aquarium *"Fish Worship-Is it Wrong? Role of the Sanctuary Advisory Council as the Liaison to the Community"*

1:30 p.m.	Committee Questions and Discussion Period
2:15 p.m.	**Gary Davis**, National Park Service, Channel Islands *National Parks in the Sea: Why Are They Different?*
2:45 p.m.	Committee Questions and Discussion Period
3:00 p.m.	**Zeke Grader**, Pacific Coast Federation of Fishermen's Associations
3:30 p.m.	Committee Questions and Discussion Period
3:45 p.m.	**Break**
4:00 p.m.	**Mark Carr**, University of California, Santa Cruz
4:30 p.m.	Committee Questions and Discussion Period
5:00 p.m.	**Tony Koslow**, CSIRO Marine Research, Hobart Tasmania Australia and NRC Senior Research Fellow Pacific Fisheries Environmental Laboratory, NMFS
5:30 p.m.	Adjourn
6:30 p.m.	Committee Dinner

Friday, May 28

8:00 a.m.	**Breakfast**
OPEN SESSION	
8:30 a.m.	**Mary Yoklavich**, NMFS, Pacific Fisheries Environmental Lab, Pacific Grove
9:00 a.m.	Committee Questions and Discussion Period
9:15 a.m.	**Richard Parrish**, NMFS, Pacific Fisheries Environmental Lab, Pacific Grove
9:45 a.m.	Committee Questions and Discussion Period
10:00 a.m.	**Break**
10:15 a.m.	**Melissa Miller-Henson**, California Ocean Resources Management Program
10:45 a.m.	Committee Questions and Discussion Period
11:00 a.m.	**Sean Hastings** and **Anne Walton**, Channel Islands National Marine Sanctuary

11:30 a.m.	**Steve Eittreim**, USGS Monterey Bay Sanctuary Program
11:45 a.m.	Committee Questions and Discussion Period
12:00 noon	**Lunch**
1:00 p.m.	**Steve Palumbi**, Report from the NCEAS Working Group
1:30 p.m.	**Richard Stoffle**, The Role of Environmental Values and Ethics in the Development of MPAs or One Positive Reason Why Folks Belong in the Ocean.
2:00 p.m.	**Jim Wilen**, Economic Issues in Marine Reserve Policy Design
2:30 p.m.	Question and Discussion Period
3:00 p.m.	Adjourn open session
6:00 p.m.	Meeting Adjourns for the Day

THE COMMITTEE ON THE EVALUATION, DESIGN, AND MONITORING OF MARINE RESERVES AND PROTECTED AREAS IN THE UNITED STATES

The Aljoya Conference Center
3920 NE 41st Street
Seattle, WA 98105

September 8-10, 1999

AGENDA

Wednesday, September 8 - Alder Conference Room

| 8:00 a.m. | **Breakfast in the Cedar Lobby** |

OPEN SESSION

| 11:30 a.m.—2:00 p.m. | **Extended Lunch Break (Pacific Dining Room)** Sponsor Presentations—William Fox, National Marine Fisheries Service Carol Bernthal, Olympic Coast National Marine Sanctuary |
| 6:00 p.m. | Meeting Adjourns for the Day |

Thursday, September 9 - Cedar Conference Room

8:00 a.m.	**Breakfast in the Cedar Lobby**
8:30 a.m.	Opening Remarks—Ed Houde, Committee Chair
9:00 a.m.	Clarence Pautzke, North Pacific Fishery Management Council
9:20 a. m.	Committee Questions and Discussion Period
9:30 a.m.	Dennis Willows, University of Washington, Friday Harbor Laboratory "Bottom-up vs. Top-down in the Context of Marine Resource Protection: What Works Best?"
9:50 a.m.	Committee Questions and Discussion Period
10:00 a.m.	Todd Pitlik, Division of Aquatic and Wildlife Resources, Guam "Marine Reserves and Coral Reef Fisheries in Guam"
10:20 a.m.	Committee Questions and Discussion Period
10:30 a.m.	**Break**
10:50 a.m.	Elliot Norse, Marine Conservation Biology Institute "MPAs: Science, Policy and Politics."
11:10 a.m.	Committee Questions and Discussion Period
11:20 a.m.	Carl Walters, University of British Columbia "Efficacy of Area Closures in Fisheries Management"
11:40 a.m.	Committee Questions and Discussion Period
12:00 p.m.	**Lunch Break in the Pacific Dining Room**
1:00 p.m.	Louise Goulet, Pacific Marine Heritage Legacy, Parks Canada "Design and Implementation of the Robson Bight Ecological Reserve: A Case Study"
1:20 p.m.	Committee Questions and Discussion Period
1:30 p.m.	Tory O'Connell, Alaska Department of Fish and Game
1:50 p.m.	Committee Questions and Discussion Period
2:00 p.m.	Stuart Ellis, Northwest Indian Fisheries Commission
2:15 p.m.	Carol Bernthal, Olympic Coast National Marine Sanctuary
2:30 p.m.	Committee Questions and Discussion Period
2:50 p.m.	Dave Fraser, Owner and Captain, F/V Muir Milach

3:10 p.m.	Committee Questions and Discussion Period
3:20 p.m.	Jim Taggart, U.S. Geological Survey "The phase out of commercial fishing and the opportunities for testing the effectiveness of marine reserves in Glacier Bay National Park"
3:40 p.m.	Committee Questions and Discussion Period
4:00 p.m.	**Break**
4:30 p.m.	PUBLIC DISCUSSION
6:00 p.m.	**Meeting Ajourns**

E

Presidential Executive Order Regarding Marine Protected Areas in the United States

THE WHITE HOUSE

Office of the Press Secretary

For Immediate Release   May 26, 2000

EXECUTIVE ORDER

MARINE PROTECTED AREAS

By the authority vested in me as President by the Constitution and the laws of the United States of America and in furtherance of the purposes of the National Marine Sanctuaries Act (16 U.S.C. 1431 et seq.), National Wildlife Refuge System Administration Act of 1966 (16 U.S.C. 668dd-ee), National Park Service Organic Act (16 U.S.C. 1 et seq.), National Historic Preservation Act (16 U.S.C. 470 et seq.), Wilderness Act (16 U.S.C. 1131 et seq.), Magnuson-Stevens Fishery Conservation and Management Act (16 U.S.C. 1801 et seq.), Coastal Zone Management Act (16 U.S.C. 1451 et seq.), Endangered Species Act of 1973 (16 U.S.C. 1531 et seq.), Marine Mammal Protection Act (16 U.S.C. 1362 et seq.),

Clean Water Act of 1977 (33 U.S.C. 1251 et seq.), National Environmental Policy Act, as amended (42 U.S.C. 4321 et seq.), Outer Continental Shelf Lands Act (42 U.S.C. 1331 et seq.), and other pertinent statutes, it is ordered as follows:

Section 1. Purpose. This Executive Order will help protect the significant natural and cultural resources within the marine environment for the benefit of present and future generations by strengthening and expanding the Nation's system of marine protected areas (MPAs). An expanded and strengthened comprehensive system of marine protected areas throughout the marine environment would enhance the conservation of our Nation's natural and cultural marine heritage and the ecologically and economically sustainable use of the marine environment for future generations. To this end, the purpose of this order is to, consistent with domestic and international law: (a) strengthen the management, protection, and conservation of existing marine protected areas and establish new or expanded MPAs; (b) develop a scientifically based, comprehensive national system of MPAs representing diverse U.S. marine ecosystems, and the Nation's natural and cultural resources; and (c) avoid causing harm to MPAs through federally conducted, approved, or funded activities.

Section 2. Definitions. For the purposes of this order:

(a) "Marine protected area" means any area of the marine environment that has been reserved by Federal, State, territorial, tribal, or local laws or regulations to provide lasting protection for part or all of the natural and cultural resources therein.

(b) "Marine environment" means those areas of coastal and ocean waters, the Great Lakes and their connecting waters, and submerged lands thereunder, over which the United States exercises jurisdiction, consistent with international law.

(c) The term "United States" includes the several States, the District of Columbia, the Commonwealth of Puerto Rico, the Virgin Islands of the United States, American Samoa, Guam, and the Commonwealth of the Northern Mariana Islands.

Section 3. MPA Establishment, Protection, and Management. Each Federal agency whose authorities provide for the establishment or management of MPAs shall take appropriate actions to enhance or expand protection of existing MPAs and establish or recommend, as appropriate, new MPAs. Agencies implementing this section shall consult with the agencies identified in subsection 4(a) of this order, consistent with existing requirements.

Section 4. National System of MPAs. (a) To the extent permitted by law and subject to the availability of appropriations, the Department of Commerce and the Department of the Interior, in consultation with the Department of Defense, the Department of State, the United States Agency for International Development, the Department of Transportation, the Environmental Protection Agency, the National Science Foundation, and other pertinent Federal agencies shall develop a national system of MPAs. They shall coordinate and share information, tools, and strategies, and provide guidance to enable and encourage the use of the following in the exercise of each agency's respective authorities to further enhance and expand protection of existing MPAs and to establish or recommend new MPAs, as appropriate:

(1) science-based identification and prioritization of natural and cultural resources for additional protection;

(2) integrated assessments of ecological linkages among MPAs, including ecological reserves in which consumptive uses of resources are prohibited, to provide synergistic benefits;

(3) a biological assessment of the minimum area where consumptive uses would be prohibited that is necessary to preserve representative habitats in different geographic areas of the marine environment;

(4) an assessment of threats and gaps in levels of protection currently afforded to natural and cultural resources, as appropriate;

(5) practical, science-based criteria and protocols for monitoring and evaluating the effectiveness of MPAs;

(6) identification of emerging threats and user conflicts affecting MPAs and appropriate, practical, and equitable management solutions, including effective enforcement strategies, to eliminate or reduce such threats and conflicts;

(7) assessment of the economic effects of the preferred management solutions; and

(8) identification of opportunities to improve linkages with, and technical assistance to, international marine protected area programs.

(b) In carrying out the requirements of section 4 of this order, the Department of Commerce and the Department of the Interior shall consult with those States that contain portions of the marine environment, the

Commonwealth of Puerto Rico, the Virgin Islands of the United States, American Samoa, Guam, and the Commonwealth of the Northern Mariana Islands, tribes, Regional Fishery Management Councils, and other entities, as appropriate, to promote coordination of Federal, State, territorial, and tribal actions to establish and manage MPAs.

(c) In carrying out the requirements of this section, the Department of Commerce and the Department of the Interior shall seek the expert advice and recommendations of non-Federal scientists, resource managers, and other interested persons and organizations through a Marine Protected Area Federal Advisory Committee. The Committee shall be established by the Department of Commerce.

(d) The Secretary of Commerce and the Secretary of the Interior shall establish and jointly manage a website for information on MPAs and Federal agency reports required by this order. They shall also publish and maintain a list of MPAs that meet the definition of MPA for the purposes of this order.

(e) The Department of Commerce's National Oceanic and Atmospheric Administration shall establish a Marine Protected Area Center to carry out, in cooperation with the Department of the Interior, the requirements of subsection 4(a) of this order, coordinate the website established pursuant to subsection 4(d) of this order, and partner with governmental and nongovernmental entities to conduct necessary research, analysis, and exploration. The goal of the MPA Center shall be, in cooperation with the Department of the Interior, to develop a framework for a national system of MPAs, and to provide Federal, State, territorial, tribal, and local governments with the information, technologies, and strategies to support the system. This national system framework and the work of the MPA Center is intended to support, not interfere with, agencies' independent exercise of their own existing authorities.

(f) To better protect beaches, coasts, and the marine environment from pollution, the Environmental Protection Agency (EPA), relying upon existing Clean Water Act authorities, shall expeditiously propose new science-based regulations, as necessary, to ensure appropriate levels of protection for the marine environment. Such regulations may include the identification of areas that warrant additional pollution protections and the enhancement of marine water quality standards. The EPA shall consult with the Federal agencies identified in subsection 4(a) of this order, States, territories, tribes, and the public in the development of such new regulations.

Section 5. Agency Responsibilities. Each Federal agency whose actions affect the natural or cultural resources that are protected by an MPA shall identify such actions. To the extent permitted by law and to the maximum extent practicable, each Federal agency, in taking such actions, shall avoid harm to the natural and cultural resources that are protected by an MPA. In implementing this section, each Federal agency shall refer to the MPAs identified under subsection 4(d) of this order.

Section 6. Accountability. Each Federal agency that is required to take actions under this order shall prepare and make public annually a concise description of actions taken by it in the previous year to implement the order, including a description of written comments by any person or organization stating that the agency has not complied with this order and a response to such comments by the agency.

Section 7. International Law. Federal agencies taking actions pursuant to this Executive Order must act in accordance with international law and with Presidential Proclamation 5928 of December 27, 1988, on the Territorial Sea of the United States of America, Presidential Proclamation 5030 of March 10, 1983, on the Exclusive Economic Zone of the United States of America, and Presidential Proclamation 7219 of September 2, 1999, on the Contiguous Zone of the United States.

Section 8. General. (a) Nothing in this order shall be construed as altering existing authorities regarding the establishment of Federal MPAs in areas of the marine environment subject to the jurisdiction and control of States, the District of Columbia, the Commonwealth of Puerto Rico, the Virgin Islands of the United States, American Samoa, Guam, the Commonwealth of the Northern Mariana Islands, and Indian tribes.

(b) This order does not diminish, affect, or abrogate Indian treaty rights or United States trust responsibilities to Indian tribes.

(c) This order does not create any right or benefit, substantive or procedural, enforceable in law or equity by a party against the United States, its agencies, its officers, or any person.

WILLIAM J. CLINTON
THE WHITE HOUSE,
May 26, 2000.

F

The IUCN Protected Area Categories System

The following notes are a brief introduction to the system. They are followed by an extract from the IUCN publication *Guidelines for Protected Area Management Categories* (IUCN, 1994).

The definition of a marine protected area (MPA) adopted by IUCN and other international and national bodies is :
Any area of intertidal or subtidal terrain, together with its overlying water and associated flora, fauna, historical and cultural features, which has been reserved by law or other effective means to protect part or all of the enclosed environment. (Kelleher and Kenchington, 1992).

The main aims of MPAs have been identified in IUCN's *Guidelines for Establishing Marine Protected Areas* (Kelleher and Kenchington, 1992) as:

* to maintain essential ecological and life support systems;
* to ensure the sustainable utilization of species and ecosystems; and
* to preserve biotic diversity.

When considering the utility of MPAs for sustaining fisheries, it would be hard to argue that the attainment of any of these fundamental aims is not essential. They are, however, general aims and they can be expanded to the following purposes, most of which are relevant to fisheries (IUCN, 1994):

* scientific research;
* wilderness protection;

- preservation of species and genetic diversity;
- maintenance of environmental services;
- protection of specific natural and cultural features;
- tourism and recreation;
- education;
- sustainable use of resources from natural ecosystems; and
- maintenance of cultural and traditional attributes.

There are several important features of the IUCN categorization scheme that it is important to note. They are:

- the basis of categorization is by **primary management objective;**
- assignment to a category is not a commentary on management effectiveness;
- the categories system is international;
- national names for protected areas of the same category vary;
- all categories are important; and
- though the primary objective of an MPA will determine the category, the MPA may contain zones which have other objectives. However, for the purpose of categorization, at least 3/4 of the MPA must be managed for the primary objective and the management of the remaining area must not conflict with that primary objective.

CATEGORY I

Strict Nature Reserve - Wilderness Area: Protected Area Managed Mainly for Science or Wilderness Protection

CATEGORY IA

Strict Nature Reserve: Protected Area Managed Mainly for Science

Definition

Area of land and/or sea possessing some outstanding or representative ecosystems, geological or physiological features and/or species, available primarily for scientific research and/or environmental monitoring.

Objectives of Management

- to preserve habitats, ecosystems and species in as undisturbed a state as possible;

- to maintain genetic resources in a dynamic and evolutionary state;
- to maintain established ecological processes;
- to safeguard structural landscape features or rock exposures;
- to secure examples of the natural environment for scientific studies, environmental monitoring and education, including baseline areas from which all avoidable access is excluded;
- to minimize disturbance by careful planning and execution of research and other approved activities; and
- to limit public access.

Guidance for Selection

- The area should be large enough to ensure the integrity of its ecosystems and to accomplish the management objectives for which it is protected.
- The area should be significantly free of direct human intervention and capable of remaining so.
- The conservation of the area's biodiversity should be achievable through protection and not require substantial active management or habitat manipulation (c.f. Category IV).

Organizational Responsibility

Ownership and control should be by the national or other level of government, acting through a professionally qualified agency, or by a private foundation, university or institution which has an established research or conservation function, or by owners working in cooperation with any of the foregoing government or private institutions. Adequate safeguards and controls relating to long-term protection should be secured before designation. International agreements over areas subject to disputed national sovereignty can provide exceptions (e.g., Antarctica).

Equivalent Category in 1978 System

Scientific Research / Strict Nature Reserve

CATEGORY IB

Wilderness Area: Protected Area Managed Mainly for Wilderness Protection

Definition

Large area of unmodified or slightly modified land, and/or sea, retaining its natural character and influence, without permanent or significant habitation, which is protected and managed so as to preserve its natural condition.

Objectives of Management

• to ensure that future generations have the opportunity to experience understanding and enjoyment of areas that have been largely undisturbed by human action over a long period of time;

• to maintain the essential natural attributes and qualities of the environment over the long term;

• to provide for public access at levels and of a type which will serve best the physical and spiritual well-being of visitors and maintain the wilderness qualities of the area for present and future generations; and

• to enable indigenous human communities living at low density and in balance with the available resources to maintain their lifestyle.

Guidance for Selection

• The area should possess high natural quality, be governed primarily by the forces of nature, with human disturbance substantially absent, and be likely to continue to display those attributes if managed as proposed.

• The area should contain significant ecological, geological, physiogeographic, or other features of scientific, educational, scenic or historic value.

• The area should offer outstanding opportunities for solitude, enjoyed once the area has been reached, by simple, quiet, non-polluting and non-intrusive means of travel (i.e., non-motorized).

• The area should be of sufficient size to make practical such preservation and use.

Organizational Responsibility

As for Sub-Category 1a.

Equivalent Category

This sub-category did not appear in the 1978 system, but has been introduced following the IUCN General Assembly Resolution (16/34) on Protection of Wilderness Resources and Values, adopted at the 1984 General Assembly in Madrid, Spain.

CATEGORY II

National Park: Protected Area Managed Mainly for Ecosystem Protection and Recreation

Definition

Natural area of land and/or sea, designated to (a) protect the ecological integrity of one or more ecosystems for present and future generations, (b) exclude exploitation or occupation inimical to the purposes of designation of the area and (c) provide a foundation for spiritual, scientific, educational, recreational and visitor opportunities, all of which must be environmentally and culturally compatible.

Objectives of Management

• to protect natural and scenic areas of national and international significance for spiritual, scientific, educational, recreational or tourist purposes;

• to perpetuate, in as natural a state as possible, representative examples of physiographic regions, biotic communities, genetic resources, and species, to provide ecological stability and diversity;

• to manage visitor use for inspirational, educational, cultural, and recreational purposes at a level which will maintain the area in a natural or near natural state;

• to eliminate and thereafter prevent exploitation or occupation inimical to the purposes of designation;

• to maintain respect for the ecological, geomorphologic, sacred or aesthetic attributes which warranted designation; and

• to take into account the needs of indigenous people, including subsistence resource use, in so far as these will not adversely affect the other objectives of management.

Guidance for Selection

• The area should contain a representative sample of major natural regions, features or scenery, where plant and animal species, habitats and geomorphological sites are of special spiritual, scientific, educational, recreational, and tourist significance.

• The area should be large enough to contain one or more entire ecosystems not materially altered by current human occupation or exploitation.

Organizational Responsibility

Ownership and management should normally be by the highest competent authority of the nation having jurisdiction over it. However, they may also be vested in another level of government, council of indigenous people, foundation or other legally established body which has dedicated the area to long-term conservation.

Equivalent Category in 1978 System

National Park.

CATEGORY III

Natural Monument: Protected Area Managed Mainly for Conservation of Specific Natural Features

Definition

Area containing one, or more, specific natural or natural/cultural features which is of outstanding or unique value because of its inherent rarity, representative or aesthetic qualities or cultural significance.

Objectives of Management

• to protect or preserve in perpetuity specific outstanding natural features because of their natural significance, unique or representational quality, and/or spiritual connotations;
• to an extent consistent with the foregoing objective, to provide opportunities for research, education, interpretation and public appreciation;
• to eliminate and thereafter prevent exploitation or occupation inimical to the purpose of designation; and
• to deliver to any resident population such benefits as are consistent with the other objectives of management.

Guidance for Selection

• The area should contain one or more features of outstanding significance (appropriate natural features include spectacular waterfalls, caves, craters, fossil beds, sand dunes and marine features, along with unique or representative fauna and flora; associated cultural features might include cave dwellings, cliff-top forts, archaeological sites, or natural sites which have heritage significance to indigenous peoples).
• The area should be large enough to protect the integrity of the feature and its immediately related surroundings.

Organizational Responsibility

Ownership and management should be by the national government or, with appropriate safeguards and controls, by another level of government, council of indigenous people, non-profit trust, corporation or, exceptionally, by a private body, provided the long-term protection of the inherent character of the area is assured before designation.

Equivalent Category in 1978 System

Natural Monument / Natural Landmark.

CATEGORY IV

Habitat/Species Management Area: Protected Area Managed Mainly for Conservation Through Management Intervention

Definition

Area of land and/or sea subject to active intervention for management purposes so as to ensure the maintenance of habitats and/or to meet the requirements of specific species.

Objectives of Management

•   to secure and maintain the habitat conditions necessary to protect significant species, groups of species, biotic communities or physical features of the environment where these require specific human manipulation for optimum management;
•   to facilitate scientific research and environmental monitoring as primary activities associated with sustainable resource management;
•   to develop limited areas for public education and appreciation of the characteristics of the habitats concerned and of the work of wildlife management;
•   to eliminate and thereafter prevent exploitation or occupation inimical to the purposes of designation; and
•   to deliver such benefits to people living within the designated area as are consistent with the other objectives of management.

Guidance for Selection

•   The area should play an important role in the protection of nature and the survival of species, (incorporating, as appropriate, breeding areas, wetlands, coral reefs, estuaries, grasslands, forests or spawning areas, including marine feeding beds).
•   The area should be one where the protection of the habitat is essential to the well-being of nationally or locally-important flora, or to resident or migratory fauna.
•   Conservation of these habitats and species should depend upon active intervention by the management authority, if necessary through habitat manipulation (c.f. Category Ia).
•   The size of the area should depend on the habitat requirements of the species to be protected and may range from relatively small to very extensive.

Organizational Responsibility

Ownership and management should be by the national government or, with appropriate safeguards and controls, by another level of government, non-profit trust, corporation, private group or individual.

Equivalent Category in 1978 System

Nature Conservation Reserve / Managed Nature Reserve / Wildlife Sanctuary.

CATEGORY V

Protected Landscape/Seascape: Protected Area Managed Mainly for Landscape/Seascape Conservation and Recreation

Definition

Area of land, with coast and sea as appropriate, where the interaction of people and nature over time has produced an area of distinct character with significant aesthetic, ecological and/or cultural value, and often with high biological diversity. Safeguarding the integrity of this traditional interaction is vital to the protection, maintenance, and evolution of such an area.

Objectives of Management

•   to maintain the harmonious interaction of nature and culture through the protection of landscape, and/or seascape and the continuation of traditional land uses, building practices and social and cultural manifestations;
•   to support lifestyles and economic activities which are in harmony with nature and the preservation of the social and cultural fabric of the communities concerned;
•   to maintain the diversity of landscape and habitat, and of associated species and ecosystems;
•   to eliminate where necessary, and thereafter prevent, land uses and activities which are inappropriate in scale and/or character;
•   to provide opportunities for public enjoyment through recreation and tourism appropriate in type and scale to the essential qualities of the areas;
•   to encourage scientific and educational activities which will contribute to the long term well-being of resident populations and to the development of public support for the environmental protection of such areas; and
•   to bring benefits to, and to contribute to the welfare of, the local community through the provision of natural products (such as forest and fisheries products) and services (such as clean water or income derived from sustainable forms of tourism).

Guidance for Selection

• The area should possess a landscape and/or coastal and island seascape of high scenic quality, with diverse associated habitats, flora and fauna along with manifestations of unique or traditional land-use patterns and social organizations as evidenced in human settlements and local customs, livelihoods, and beliefs.
• The area should provide opportunities for public enjoyment through recreation and tourism within its normal lifestyle and economic activities.

Organizational Responsibility

The area may be owned by a public authority, but is more likely to comprise a mosaic of private and public ownerships operating a variety of management regimes. These regimes should be subject to a degree of planning or other control and supported, where appropriate, by public funding and other incentives, to ensure that the quality of the landscape/seascape and the relevant local customs and beliefs are maintained in the long term.

Equivalent Category in 1978 System

Protected Landscape

CATEGORY VI

Managed Resource Protected Area: Protected Area Managed Mainly for the Sustainable Use of Natural Ecosystems

Definition

Area containing predominantly unmodified natural systems, managed to ensure long term protection and maintenance of biological diversity, while providing at the same time a sustainable flow of natural products and services to meet community needs.

Objectives of Management

• to protect and maintain the biological diversity and other natural values of the area in the long term;
• to promote sound management practices for sustainable production purposes;
• to protect the natural resource base from being alienated for other land-use purposes that would be detrimental to the area's biological diversity; and
• to contribute to regional and national development.

Guidance for Selection

•   The area should be at least two-thirds in a natural condition, although it may also contain limited areas of modified ecosystems; large commercial plantations would *not* be appropriate for inclusion.
•   The area should be large enough to absorb sustainable resource uses without detriment to its overall long-term natural values.

Organizational Responsibility

Management should be undertaken by public bodies with a unambiguous remit for conservation, and carried out in partnership with the local community; or management may be provided through local custom supported and advised by governmental or non-governmental agencies. Ownership may be by the national or other level of government, the community, private individuals, or a combination of these.

Equivalent Category in 1978 System

This category does not correspond directly with any of those in the 1978 system, although it is likely to include some areas previously classified as 'Resource Reserves', 'Natural Biotic Areas / Anthropological Reserves' and 'Multiple Use Management Areas / Managed Resource Areas.'

SOURCE: IUCN, 1994; Kelleher and Kenchington, 1992.

G

Description of Studies Estimating Marine Reserve Area Requirements

(Amended from Roberts and Hawkins, in press, and reproduced with permission)

OBJECTIVE: ETHICS

Ballantine, 1997:

Argues for a target of 10% of all of the marine habitats in New Zealand to be protected. The key principle at stake is that we should not fish everywhere. Some areas should be set aside as refuges from exploitation for ethical reasons. Ten percent, Ballantine says, "has a long traditional use as a figure that signifies importance without serious hurt." It contrasts favorably with the 90% left open to exploitation and is conservative compared to the protected land area of New Zealand. However, he accepts that it represents a call to arms for conservation rather than being scientifically-based.

OBJECTIVE: RISK MINIMIZATION

Lauck et al., 1998:

Examined the combined effects of variation in stock productivity, and errors in estimating mortality and population size, on the probability of managers successfully maintaining populations above target levels. In a simple model showed that, in the face of uncertainty in fishing mortality, reserves covering between 31 and 70% of fishing grounds would be needed to maintain populations above 60% of their unexploited size (argued to be an economic optimum) over a 40 year time horizon. The area of reserve required increased with fishing intensity. Furthermore, the greater the uncertainty in fishing mortality (which is equivalent to decreasing management control), the larger the reserves required.

Roughgarden, 1998:

Recommended maintaining exploited populations at 75% of their unexploited size in order to avoid recruitment overfishing.

Guénette et al., 2000:

Used a spatially explicit model to examine whether reserves could have prevented the collapse in 1992 of the migratory northern cod (*Gadus morhua*) population off eastern Canada. Found that, in the absence of other management measures, reserves covering 80% of the area would have been necessary, but that with temporal closures to trawls and gill nets, reserves covering 20% of the area would have been adequate.

Mangel, 2000:

Looked at the use of reserves as a tool to maintain fish populations above target levels. Found that if a stock was initially heavily fished (i.e., starts at 35% of its unfished size) reserves of 20 and 30% of the management area guaranteed persistence above this level for 20 and 100 years, respectively. The greater the level of population desired, the longer the planning horizon, and the higher the degree of variability in fishing mortality (= less control over fishing), the larger reserves are required to maintain target populations. Reserves increased cumulative yields from the fishery when populations were initially heavily exploited.

Goodyear, 1993:

Used fishery models to estimate that maintaining fish populations above 20% of their unexploited size would avoid recruitment overfishing.

Mace and Sissenwine, 1993:

For 91 fish populations (representing 27 species) in North America and Europe, calculated that the average threshold replacement stock size corresponds to a 20% spawning potential ratio (one fifth of unexploited population size). Maintaining at least a 30% spawning potential ratio would avoid recruitment overfishing for 80% of these species; therefore, a 35% spawning potential ratio would be a conservative management target. However, safe minimum population levels ranged as high as 70% for some species.

Mace, 1994:

Argued that, where the nature of the relationship between population size and recruitment is unknown, a precautionary approach would be to aim to maintain populations above 40% of their unexploited size.

Sumaila, 1998:

Used a bioeconomic model to examine effects of different reserve areas on economic yields from the Barents Sea cod fishery. Reserves reduced economic yield from the fishery but increased cod population size. The system was also modeled with an ecological shock - a ten year period of recruitment failure. Reserves supported populations through this recruitment failure and were found to be bioeconomically beneficial when there were moderate levels of movement of cod from reserves to fishing grounds (40 to 60% of cod leave reserve in a year). This allowed reserve benefits to be captured by the fishery. The largest reserves modeled, covering 70% of the management area, offered the greatest future security for stocks, but had the highest cost in terms of current yields. How large reserves should be depends on the degree to which populations are subject to external shocks, and the degree of risk managers are willing to accept. In general, reserves covering 30 to 50% of the area provided significant protection for stocks without greatly reducing current economic benefits.

Man et al., 1995:

Modeled the persistence of an exploited metapopulation distributed across a series of habitat patches. Reserves (protected patches) became highly beneficial to population persistence as the local extinction rate in patches increased (due to increasing fishing intensities). This is because reserves provided a source of offspring to replenish fished out patches. Reserves became beneficial as exploitation rates increased, reaching a maximum of 50% of the patches protected at the highest levels of fishing. However, over a wide range of fishing intensities, optimal reserve fractions ranged between 20 and 40%.

OBJECTIVES: RISK MINIMIZATION AND BYCATCH AVOIDANCE

Soh et al., 1998:

Modeled the effects of closing hotspot areas for catches of two species of rockfish in the Gulf of Alaska. The fishery for these species is unselective and currently there are high levels of discards of over-quota fish, ranging from 15 to over 60% of catches. Three areas of reserves were simulated, covering approximately 4, 9 and 16% of the trawlable shelf area of the region. Because reserves allowed all catches to be landed, rather than fishers having to discard fish, none of the reserve areas resulted in reduced catches. Reserves played a key role in increasing biomass of both species over a 20 year time horizon, whereas without reserves, biomass declined. The authors concluded that placing reserves in hotspots of adult fish biomass would enable even the smallest areas simulated to significantly improve on current management.

OBJECTIVES: RISK MINIMIZATION AND YIELD MAXIMIZATION

Foran and Fujita, 1999:

Modeled the value of reserves on rebuilding egg output by stocks of Pacific Ocean perch (*Sebastes alutus*), and catches, under optimistic and pessimistic assumptions of recruitment. Found that the benefits of reserves were sensitive to levels of recruitment. For example, a 10% reserve system would decrease long-term catches by 8% if recruitment were good, while the same reserve would increase catches by 15% if recruitment were poor. As the fraction protected increased, so fishing rates outside reserves had to be increased to maintain yields. The maximum long-term catch was from a reserve area of 25% and a moderately heavy level of fishing outside. The highest catch levels can be maintained using a range of reserve sizes provided fishing effort outside can be adjusted to appropriate levels. However, reserves increased the resilience of the stock to higher levels of fishing and therefore provide a risk averse management approach.

Guénette and Pitcher, 1999:

Used a dynamic model, which included weight-fecundity and stock-recruitment relationships to examine the effects of reserves on cod (*Gadus morhua*). Found that reserves do not increase yields until cod are exploited at higher levels than necessary to achieve maximum sustainable yield. At higher fishing intensities, reserves prevented collapse in catch, with 30% reserves maintaining the highest yields of the four reserve areas modeled (10, 30, 50 and 70%). Larger reserves (> 30% protected) provided more robust biomass of spawning fish and reduced the number of years with poor recruitment compared to a no reserve regime. Increasing transfer rates of fish from reserves to fishing grounds decreased the benefits from reserves. However, even for highly mobile fish, reserves should be able to maintain higher spawning stocks than without them.

OBJECTIVE: YIELD MAXIMIZATION

Pezzey et al., in press:

In a bioeconomic model showed that the reserve area that maximized catches in coral reef fisheries varied between 0 and 50% of the total area, depending on the intensity of fishing outside reserves. As fishing intensity increases, so greater fractions of the fishing grounds must be protected to sustain catches. They calculated that reserves covering 21%, 36% and 40% would be required to sustain yields in the fisheries of Belize, St. Lucia and Jamaica, representing a gradient from moderate to intensive exploitation.

Sladek Nowlis and Roberts, 1997, 1999:

Using a single-species model, applied to four different species, showed that the fraction of a management area required in reserves depends on intensity of exploitation. Reserves were only effective in increasing catches when species were overfished. As fishing intensity increases, larger and larger reserves are required to sustain catches. In the most intensively exploited areas of the Caribbean, reserves covering 75-80% would be needed to maximize catches. However, at more moderate fishing intensities, reserves covering 40% of the management area would offer major benefits to yields.

Sladek Nowlis, 2000:

Modeled the effects of reserves on catches of the Caribbean white grunt (*Haemulon plumieri*). At moderate fishing intensities (20% of fishery recruited individuals removed per year) catches peaked with reserves covering 30% of the management area.

Sladek Nowlis and Yoklavich, 1998:

Used a population model to examine the potential for reserves to enhance catches of a Pacific rockfish, the Boccacio (*Sebastes paucispinis*). They found that reserves could produce moderate to great enhancements in catch depending on how overfished the species was to begin with. Optimal reserve areas, those producing the greatest long term catches, ranged from roughly 20 to 27% of the management area as fishing intensities grew.

Holland and Brazee, 1996:

Simulated the effects of reserves on catches from the red snapper (*Lutjanus campechanus*) fishery in the Gulf of Mexico. Found that reserves would not benefit catches until the species was overfished. For a range of heavy exploitation rates, optimal reserve areas (those that maximized catches) increased from 15 to 29% of the area as fishing pressure increased. However, in economic terms of net present value, optimal reserve areas were reduced from these values as the rate of discounting of the future increased (in other words as the relative value afforded to present compared to future catches grows).

Hannesson, 1998:

Used a bioeconomic model to examine effects of reserves on spawning stock size, catches and costs of fishing for a mobile species like the cod (*Gadus morhua*). Assumed open access fishing outside reserves and found that reserves would have to be very large (70-80% of the management area) in order to produce catches and spawning stock levels equivalent to those of an optimally controlled fishery (one where stock size is held at 60% of the unexploited level). However, optimal control is an unrealizable economic abstraction and, compared to open access, reserves fared well. When covering between 50 and 80% of the area they produced increases in spawning stocks of 40-130%. Catches were greater than open access over a range of 10-80% of the area protected. The area that needs to be protected reduces when controls on fishing are implemented in remaining fishing grounds. However, reserves increased the costs of fishing and tended to promote overcapacity. The model ignored possible increases in catch from increased reproduction by the stock.

Polacheck, 1990:

Used a yield per recruit model for Georges Bank cod (*Gadus morhua*) to examine reserve effects on spawning stock biomass and yield in relation to reserve area, fishing pressure and rate of movement of fish from reserves to fishing grounds. Reserves were very effective at increasing spawning stock biomass. However, they decreased catches unless there were moderate rates of movement of fish from reserves to fishing grounds (although the model did not consider possible enhancements in catch that might be provided by increased reproduction by protected stocks). Reserves became more effective as fishing intensities increased, and the area of reserve needed to increase catch grew as the mobility of the fish increased. For transfer rates from reserve to fishing grounds of 50% of the population per year, reserve areas of between 10 and 40% of the fishing grounds increased catches, the area needed rising over this range as fishing intensities increased.

DeMartini, 1993:

Used a yield per recruit model to examine effects of reserves on catches of fish on Pacific coral reefs. Reserves substantially increased spawning stock biomass for three model fish species with differing levels of mobility. Spawning stock increases were greatest for the least mobile species, and reserves became more beneficial as fishing intensities increased. However, reserves almost always decreased yield per recruit. Nevertheless, increases in spawning stock biomass reduce risk of over-exploitation, and reserves ranging from 20 to 50% of the management area would offer significant levels of insurance against overfishing, although at increasing cost to present catches. Like Polacheck (1990), DeMartini ignored the possible benefits from increased reproduction by protected stocks. If included, reserves could potentially have increased catches (see Sladek Nowlis and Roberts 1997, 1999).

Hastings and Botsford, 1999:

Found that, for a wide range of biological conditions, marine reserves could offer equivalent yields to conventional fishery management tools. For species that reproduce over long lifespans, the fraction of area that needs to be protected as reserves is smaller than the fraction of the adult population that needs to be protected under conventional management. This is because animals can reproduce over longer periods in reserves than fishing grounds. For example, maintaining reproductive output at 35% of the unexploited level might require less than 35% of the area in reserves.

Botsford et al., 1999:

Modeled the effects of reserves on catches of California red sea urchins (*Strongylocentrotus franciscanus*). They showed that reserves would benefit catches where the slope of the stock versus recruitment curve is shallow (i.e., the species is vulnerable to recruitment overfishing). By contrast, if the slope is steep, and the species is therefore resilient to recruitment overfishing, reserves would reduce catches (although still increasing spawning stocks). However, the value of the slope is uncertain for most fished species, including this urchin. They found that, over the range of vulnerability where reserves would increase catches, the catch-maximizing fraction of the management area in reserves varied from 8 to 33%. For the most probable level of vulnerability for the sea urchin, they concluded that reserves covering 17% of the coast could increase long-term equilibrium catches by 18%.

Attwood and Bennett, 1995:

Modeled the effects of reserves on catches of three species of surf zone fish that are targeted by recreational anglers. Reserves would increase catches for two of the species, while reducing risk of recruitment overfishing of the third by increasing spawning stocks. Modeled catches of Galjoen (*Dichistius capensis*) peaked at 65% of the fishing grounds in reserves, while those for blacktail (*Diplodus sargus*) peaked at around 25-30% of the coast protected. The results suggested a combined management strategy would be successful for the three species, with one third of the area protected distributed into reserves between 7 and 22km long across the coast of South Africa.

Quinn et al., 1993:

Used a population model to explore the role of reserves in managing the fishery for the red sea urchin (*Stronglyocentrotus franciscanus*) in California. This species is subject to strong Allee effects at reproduction and at recruitment. They require high adult densities for successful fertilization of eggs, and juveniles recruit to areas of high adult density and survive best under an adult 'spine canopy.' The authors simulated the effects of reserves on population sizes and catch rates for no reserves and three reserve areas: 17, 33 and 50% of the coast. Population sizes and sustained catches were greatest with 50% of the coast protected for all except the lightest level of fishing examined. This result was partly due to the spacing of reserves in relation to dispersal distance of the sea urchins. At the lowest fraction of the coast protected, reserves were too far apart for offspring to disperse from one to another.

Daan, 1993:

Simulated the effects of creating reserves in the North Sea on the fishing mortality of cod (*Gadus morhua*). Found that creating reserves covering 10% of the area would lead to reduction of mortality of only 5% at the lowest transfer rate of cod from reserves to fishing grounds. Protecting 25% could reduce mortality by 10-14%. However, cod were assumed to be homogeneously distributed across the region as was fishing effort. A more realistic simulation would probably have found greater benefits from protecting the same fractions of the area but in places where cod are more aggregated and catch higher.

OBJECTIVE: BIODIVERSITY REPRESENTATION

Turpie et al., in press:

Divided the South African coast into fifty-two 50km sections to explore designs for systems of marine reserves that would represent all species of marine fish present, and all biogeographic areas. Analyses of complementarity were used to design the most space-efficient systems of reserves. A system covering 10% of the coast could be designed that would represent 97.5% of the species. However, this would not represent 15 species endemic to South Africa. A reserve system covering 29% of the coast would represent all of the species. Representing all species in the core regions of their ranges, a commonly stated conservation goal, would require 36% of the coast to be protected.

Bustamante et al., 1999:

Developed a design for a representative system of fully-protected zones for coastal habitats in the Galapagos Marine Reserve. This reserve covers the entire archipelago. Their objectives were to protect all of the 'visiting sites' in the archipelago, areas of high biological importance, and to represent all the different coastal habitat types in each of the five biogeographic zones encompassed by the islands. To achieve this, they calculated it would be necessary to protect 36% of the coastline from fishing.

Halfpenny and Roberts, in review:

Designed a reserve system for the continental shelf seas of north-western Europe with the aim to represent all habitats and biogeographic regions present, and to replicate them in different reserves. Two systems covering 10% of the region were designed and were successful in achieving sufficient replication for most, but not all of the biogeographic regions and habitats.

OBJECTIVE: MAINTENANCE OF GENETIC VARIATION

Trexler and Travis, 2000:

Modeled the effects of fully-protected reserves to prevent or reverse undesirable selective effects of fishing, and promote genetic diversity. Found that, under the most likely selective regimes, a reserve covering just 1% of the management area would have marked conservation benefits. Benefits increased rapidly with the proportion of the area in reserves. A 10% reserve decreased directional selection by 60%, while a 20% reserve would eliminate the selective effects of fishing from the population entirely.

OBJECTIVE: INCREASE CONNECTIVITY AMONG RESERVES

Roberts, in review a:

Used a simple model in which reserve size and the fraction of the management area covered by reserves were varied to explore connectivity among reserves. Connectivity rapidly increased (= decreasing inter-reserve distances) as the proportion protected increased. For any given reserve proportion, connectivity also increased as the size of individual reserves was decreased. Connectivity increases were asymptotic, with the greatest decreases in inter-reserve distance manifested over the range of 5-30% of the management area protected, with reserves getting 76% closer to each other over this range of protection. He also examined connectivity as the 'target size' of reserves for dispersing propagules, expressed as the number of degrees of horizon covered by reserves. Target size increased steeply as the proportion of the management area protected grows, and was four times greater at 30% of the area in reserves compared to 5%.

OBJECTIVE: MAINTENANCE OF UNDISTURBED HABITAT

Allison et al., in review:

Looked at the effect of natural and human catastrophes on coastal ecosystems. Calculated that, if our aim is to protect a certain proportion of habitats in an undisturbed state, we must protect a larger fraction of the area. How much larger depends on the spatial extent of disturbance events, their frequency and rate of recovery of habitats. The more frequent a disturbance, and the longer the recovery time, the larger the fraction of a management area that must be protected in order to meet conservation targets.

Index

W

Z